Cambridge Studies in Ecology presents balanced, comprehensive, up-to-date, and critical reviews of selected topics within ecology, both botanical and zoological. The Series is aimed at advanced final-year undergraduates, graduate students, researchers, and university teachers, as well as ecologists in industry and government research.

It encompasses a wide range of approaches and spatial, temporal, and taxonomic scales in ecology, including quantitative, theoretical, population, community, ecosystem, historical, experimental, behavioural and evolutionary studies. The emphasis throughout is on ecology related to the real world of plants and animals in the field rather than on purely theoretical abstractions and mathematical models. Some books in the Series attempt to challenge existing ecological paradigms and present new concepts, empirical or theoretic models, and testable hypotheses. Others attempt to explore new approaches and present syntheses on topics of considerable importance ecologically which cut across the conventional but artificial boundaries within the science of ecology.

Bats: a community perspective

CAMBRIDGE STUDIES IN ECOLOGY

Bats

a community perspective

JAMES S. FINDLEY

Museum of Southwestern Biology
Department of Biology
University of New Mexico
Albuquerque

Published by the Press Syndicate of the University of Cambridge
The Pitt Building, Trumpington Street, Cambridge CB2 1RP
40 West 20th Street, New York, NY 10011–4211, USA
10 Stamford Road, Oakleigh, Victoria 3166, Australia

First published 1993

Printed in Great Britain at the University Press, Cambridge

A catalogue record for this book is available from the British Library

Library of Congress cataloguing in publication data

Findley, James S. (James Smith)
    Bats : a community perspective / by James S. Findley.
       p.   cm. – (Cambridge studies in ecology)
    Includes bibliographical references (p.   ) and indexes.
    ISBN 0-521-38054-5 (hardback)
    1. Bats.   2. Bats – Ecology.   3. Animal communities.   I. Title.
    II. Series.
    QL737.C5F56   1993
    599.4'045247 – dc20   92-25526   CIP

ISBN 0 521 38054 5 hardback

SE

Contents

Preface

When I was first approached, by John Birks, concerning the possibility of writing a book on bat communities, my initial reaction was 'What is there to write about?' Like many students of bat biology I have investigated assemblages of bat species, and have been intrigued by the species–rich faunas of the tropical, and especially neotropical, regions. Especially in the neotropics, the diversity of morphologies and trophic orientations seen among bats provides a strong incentive to pursue natural history investigations. But because of the difficulties in finding out what bats are doing, the hope of comprehending a local bat community as a biotically integrated whole has always seemed illusory.

Nonetheless, I concluded that by assembling the available information some generalities might emerge that would point the way to more incisive questions, and suggest future avenues of research.

But bats are not birds. There is no rich world literature comparable to that available to the avian ecologist. Unlike many groups of birds, however, bats offer some life history features which may make them, as a group, of especial interest to the community ecologist. And it may be that bats, as a relatively small group of predominately tropical animals, embody some patterns which stand forth more clearly simply because of these qualities.

This is a book about some aspects of bat biology in the context of community ecology questions. It is not a book about community ecology; we chiroptologists may never be completely ready for that. But there *is* something to write about. An overview of world bat faunas points suggestively to the conclusion that geographic and historic factors, not routinely considered by community ecologists, have exercised substantial control over the evolution and structure of local communities.

Acknowledgments

Much of what I know about bats I have learned in conjunction with other people. Many of my students and colleagues at the University of New Mexico have shared with me the excitement of studying bats, both in the laboratory and in the field. Interactions with Scott Altenbach, Hal Black, Mike Bogan, Jay Druecker, Trish Freeman, Ken Geluso, Gary Graham, Clyde Jones, Cliff Lemen, Ken Mortensen, Mike Schum, and Don Wilson, among many others, have helped shape my understanding of bat biology. Colleagues from the ongoing North American Bat Research Symposia, especially Brock Fenton, Ted Fleming, Karl Koopman, and James Dale Smith, have deepened my insight into the natural history of bats.

I have spent a lot of time in museums looking at bats, and I am especially grateful to Gorden Corbet, John Hill, Karl Koopman, Syd Anderson, Guy Musser, Charles Handley, Don Wilson, and Henry Setzer for allowing me to study material under their care at the British Museum (Natural History), the American Museum of Natural History, and the US National Museum of Natural History.

A mammalogist venturing into community ecology requires guidance, and I have received a substantial amount from Mike Rosenzweig, John Wiens, and Jim Brown. Discussions and debate with these colleagues have broadened my horizons substantially. I absolve them all from responsibility for those of my ideas promulgated herein which they find embarrassing, and I freely acknowledge their contributions to any which they may like.

Scott Altenbach made available a selection of his excellent photographs. In my opinion, Scott is the first and best American photographer of bats in flight. Many of the animals figured were captured in 1974 when he and I shared some weeks in that most magical of Costa Rican rain forests, Finca La Selva.

John Birks, James Brown, and Robert Paine read and commented upon the entire manuscript, and Howard Snell reviewed the remarks on reptiles in Chapter 1. Their insightful remarks and suggestions were instrumental in improving the clarity and balance of the work. I am greatly indebted to, and thank, all of these reviewers. They are in no way responsible for surviving infelicities in the pages that follow; those are my sole responsibility.

Writing a book takes a lot of time away from the normal pursuits and responsibilities of life. I am indebted to Don Duszynski, chairman of the Department of Biology at the University of New Mexico, for his tolerance of my frequently curmudgeonly disposition, and lack of enthusiasm for, not to mention lack of attendance at, a variety of meetings which were undoubtedly important but which, perhaps for the better, happened without me. Without his understanding, my undertaking would have been much more difficult. The initial work on the book took place while I enjoyed a sabbatical leave from my duties at the University.

I wrote most of this book at my home in Corrales, New Mexico. This naturally tranquil place was made more so by the efforts of my wife, Tommie Findley, who managed to keep a variety of children, animals, and bucolic duties from intruding into my study. She may have wondered, from time to time, why I spent so many hours on a project with such modest commercial prospects. But as an artist, she understood very well that each of us must do our thing.

1 · *The study of bat communities*

Communities are assemblages of organisms living together in the same place. Studying them is rewarding because it leads to an understanding of the way organisms of different kinds affect each other and of how they have adapted to each other's presence. Community study also provides insight into the likely consequences of alterations to the system through removal or addition of species and manipulation of environmental factors. These insights remain our only reliable guide to the formulation of practical management plans.

Communities are delineated arbitrarily, depending upon the preferences and goals of the investigator. Community ecologists often focus on a taxonomically defined subset of the local organisms, such as the birds, rodents, or grasses. Sometimes the emphasis is on a guild (Root, 1967), a set of local species using the same resource, such as the rodents, birds, and ants feeding on the seeds of desert plants. A taxon–guild comprises of a guild within a taxon, such as cavity-nesting birds, seed–eating rodents, or planktivorous fishes.

Bats are highly mobile organisms, and the species that may appear at one site reflect the fauna of a much larger area. Few of the species recorded at a single location are in permanent residence at that site. Thus the use of the term 'community' as applied to bats may refer to those occurring in a fairly large area of more or less uniform habitat, and as suggested in Chapters 6 and 7, this may sometimes include a high percentage of a regional fauna. Joseph Grinnell (1922) noted '. . . it is only a matter of time . . . until the list of California birds will be identical with that in North America as a whole.' To a somewhat more limited extent, this kind of expectation is not unreasonable for community lists of bats in many tropical regions.

Views on the reality of communities and their validity as subjects for study have varied widely. Frederic Clements (1916) looked upon communities as superorganisms, their component individuals and species playing roles analogous to those of cells and organs in traditional organisms. He viewed the establishment and development of communities through succession as comparable to the birth and ontogeny of individuals, and considered that each community had a characteristic composition that did not change until the boundary of the community was reached. By contrast, Gleason (1926) considered each species to be completely independent, the community being a random assemblage of plants and animals, and Whittaker (1956) showed that the boundaries between plant communities were far from clear-cut. The Hutchinson–MacArthur paradigm (e.g., Hutchinson, 1959; MacArthur and Levins, 1967; MacArthur and Wilson, 1967; MacArthur, 1972a) led to a view of communities as entities in equilibrium. Many subsequent workers saw the equilibrium as being produced by biotic interactions, and viewed community evolution as a deterministic process. Ecosystem ecology, as conceptualized by Odum (1967), emphasized interactions within and among trophic levels in communities as well as within the wider biosphere. Failure to detect the patterns predicted by community theory (e.g., Wiens and Rotenberry, 1980) or failure to demonstrate that alleged patterns were other than results best explained by random processes, led to rigorous questioning of the deterministic view (e.g., Wiens, 1977; Simberloff, 1983). From this re-examination emerged what has been heralded as a 'New Ecology' (Price, et al. 1984), a hallmark of which is a pluralistic view of communities. Schoener (1986) envisioned a suite of 'primitive' organismic and environmental axes ordinating such factors as body size, homeostatic ability, severity of physical factors, spatial fragmentation, and long-term climatic variation, and a suite of 'derived' axes including community characteristics such as species abundances, variation in population size, and relative importance of predation versus competition. Placement of a community on a derived axis would be some function of the contribution of the primitive axes to the derived one. Such a model, foretold as long ago as 1927 by Charles Elton, suggests the existence of different kinds of communities, depending upon defining conditions (i.e., the placement of the community on the primitive axes). Despite the implication of massive increases in the difficulty of formulating useful predictive hypotheses, Schoener (1986) expressed cautious optimism at the prospect of a manageable community theory. At the same time, the growing awareness by ecologists that biological populations may be subject to unpredictable fluctuations as a result of slight changes in initial conditions (Poole, 1989) argues strongly for the importance of continued long-term empirical community studies.

In an effort to avoid the complex implications of the term 'community,' many ecologists refer to the group of species under study as an assemblage. However this term, too, carries a certain theoretical baggage: it may imply randomness in the coming together of the various organisms, or, alternatively, the working of 'assembly rules,' in the sense of Diamond (1975). In this work, the term 'bat community' refers to all the species occurring together at one geographical locality. No processes or theoretical contexts are implied.

Why study communities?

In this climate of evolving concepts it is useful to clarify the rationale for studying assemblages of coexisting organisms.

All organisms affect each other to some degree. The most intimate associations may be represented by eukaryotic cells and their putative endosymbionts, the mitochondria and plastids, inseparable partners in the structure of what biologists regard as one living unit, yet in origin perhaps separate species. Obviously originating as separate kinds of organisms are the algal and fungal partners in lichens, which have a substantial dependence on the relationship. Slightly less interdependent are the fungi and green plants associated in mycorrhizae. Here organisms which are recognizably distinct depend on one another for normal life, yet may exist apart. Such obligate or near-obligate relationships grade into a variety of coevolved systems where the partners, though living separate lives, are mutually interdependent, such as the yucca–yucca moth pollination partnership. Parasite–host relationships often involve unilateral dependencies such as that of schistosome worms which require two hosts to complete their life cycle. A large number of biotic interrelationships are less rigidly structured, including most plant–pollinator and many host–parasite systems. Organisms comprising local communities and more widespread ecosystems are clearly dependent upon one another in a variety of ways, but the dependencies are usually less specific, and a fair amount of substitution and elimination may be possible without substantially altering the gross configuration of the system. There is no logical place in this spectrum to stop considering the assemblage as an interrelated whole. To understand an organism, one must understand its biotic context as well as its content.

Why study taxonomically defined communities?

Why should it be especially meaningful to study communities of bats, birds, or lizards? Population ecologists need no excuse to study monospecific populations. Increased understanding of population processes and interrelationships affecting individual species make such an effort richly rewarding. Population biologists routinely note interactions with predators, prey, and competitors, expecially the more important ones. If a close relative of the species of interest occurs in the area, it is likely that interactions with that species will be of special interest because close relatives are the most likely of any organisms in the local community to be similar to the study organism in requirements and to stimulate adaptations on its part. Additional relatives are interesting also in proportion to their degree of relatedness to the species of interest. More distant relatives, ones with a considerable history of adaptation else-where, might impinge less, but would still be more likely to be an important influence than organisms belonging to other genera or fami-lies. Viewed in this way, an assemblage of phylogenetic relatives, existing sympatrically in a local set of habitats, is a unit of special ecological interest.

Of course very distantly related taxa may overlap in requirements and emerge as potential competitors. Marquet (1990) reviewed this situation, and urged the importance of emphasizing the guild in studies of the role of competition in structuring communities.

Study of a taxon-community should illuminate processes and mechan-isms, such as competition and resource allocation, which allow species to coexist, especially if the suite of species differentiated in the region of study, that is, if it is autochthonous. Here, if anywhere, expected community processes should be demonstrable. Moreover, since most life is considered to be monophyletic in origin, there is a sense in which the monophyletic community is a paradigm of the earliest assemblages of living organisms, and its study may be most critical to an understanding of community evolution.

Why study bat communities?

It is thus reasonable for the community ecologist to focus attention on taxonomic groupings of organisms, beyond the pragmatic reason that such a unit is easy to identify. Are there good reasons for selecting bat communties when, as we will see, bats are not the easiest organisms with which to work? There are, indeed, at least two major attributes that commend the Chiroptera to the attention of community ecologists.

First, there are a number of autochthonous, species-rich bat assem-blages available for study. Outstanding in this regard is the magnificent fauna of phyllostomoid bats (families Phyllostomidae, Mormoopidae, and Noctilionidae) of the Neotropical region. This monophyletic assemblage of approximately 150 species seems certain to have originated and to have done all its speciating and subsequent differentiation in the neotropics. Encompassed within the confines of this group is almost the full extent of trophic diversity displayed by bats. Included are hovering gleaners of arthropods and small vertebrates, frugivores, vampires, bird and bat predators, a frog specialist, piscivores, nectar and pollen feeders, and aerial insectivores. In single localities may be encountered suites of sister taxa, of related genera within the same subfamily, members of different subfamilies, and members of different families within the phyllostomoid complex. At least 53 species in this taxon have been taken at a single site in Venezuela (Handley, 1976). All levels of interaction may be observed, from the exploitative or interference competition of newly speciated sister taxa, through predation, to casual interactions between distantly related members of different feeding, foraging and roosting guilds. On the surface, at least, these bats seem to be tantalizing subjects for community study.

Second, there is good reason to believe that bats live in a truly MacArthurian world. Every facet of their biology suggests that they are occupants of stable, predictable habitats, that they maintain stable populations close to the carrying capacity of the environment, and that their communities may exemplify those resource-limited, competition-based assemblages envisioned in a simpler age of community theory. In the ensuing section support for this contention is provided.

Bats as unique ecological beings

Some of the features which make bats distinctive also place them in an ecological and evolutionary position unexpected for animals of their size. Bats, though small, are long-lived, have low fecundity and high survivorship, a relatively long period of infant dependency, relatively advanced age at sexual maturity, and, probably because of aerodynamic requirements, they are relatively invariable morphologically as adults. They are, as a group, K-strategists, and they seem best suited to life in stable communities in equilibrium. Here some of these features are discussed in more detail.

Longevity

Bats live a long time (Fig. 1.1). The current longevity record for a free-living bat is 31 years, reported for *Myotis lucifugus* from Canada (Keen

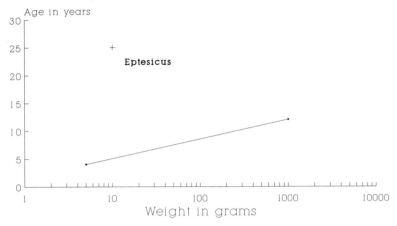

Age in years

Fig. 1.1. Regression line showing the usual relationship between weight and maximum life-span in small mammals (Sacher, 1959). A 10-g bat is plotted for contrast.

and Hitchcock, 1980), and the mean of 41 longevity records summarized by Tuttle and Stevenson (1982) is 14.6 years. The regression relating life span to weight in kilograms (W) in mammals in general (Sacher, 1959) is

$$\text{Life span} = 11.6 \ W^{0.20}.$$

It is thus predicted that an 8 g *Myotis* will attain 4.5 years. But there are many records of banded temperate-zone bats being recovered after 20 or more years. *Mean* life expectancy for seven species reported by Gaisler (1979) is 3.3 years, with extremes of 1.8 and 4.5, but this may be an underestimate because of the difficulty of recovering marked bats. Temperate-zone bats, which are mostly of the families Vespertilionidae and Rhinolophidae, almost universally hibernate, and the suggestion has been made that their longevity is a result of the prolonged periods of 'down time' that they experience. Many species of *Myotis* become dormant in September and emerge again in April, and are active only about 40% of the year. By this reasoning, a 20-year-old bat would have completed 8 years of metabolic activity, still well above the expectation for such a small mammal. Moreover tropical bats which do not hibernate or experience periods of dormancy seem to be unusually long-lived as well. Fleming's (1988) survivorship curve for the phyllostomid *Carollia perspicillata* in Costa Rica levels out at about 10.5 years. Several individuals of the neotropical vespertilionid *Myotis nigricans* have been recovered 7 years after banding in Panama, and a phyllostomid, *Artibeus jamaicensis*, attained 18 years (Wilson and Tyson, 1970). Fleming (1988) estimated the life expectancy of newborn *Carollia* at 2.56 years, 5.6 times that of a comparably sized terrestrial mammal. Herreid (1964) noted that tropical bats do not appear to have shorter life spans than temperate ones, and concluded that the extended life span of the chiropterans is not to be attributed to intervals of lowered metabolic activity. What is its explanation? Obviously bats suffer a lower mortality than many other vertebrates.

Mortality and survivorship

Once bats reach adulthood (after their first winter, if they are hibernators), they have a very good chance, averaging from 50% to 80%, of surviving through each ensuing year (Hill and Smith, 1984). Davis (1966) found that the 5.5 g *Pipistrellus subflavus* has a mean annual adult survival rate of 55% (females) and 70% (males). Seventy to 80% of banded *Plecotus townsendii* returned each year over a 3-year study (Pearson, Koford and Pearson, 1952). Bradbury and Vehrencamp (1977) determined survival rates of 54%, 78%, 79%, and 80% in four small, neotropical sac-winged bats (Emballonuridae). Heideman and Heaney (1989) recorded 60–80% survivals for four 20–90 g rain-forest megabats (Pteropidae) in the Philippines. Fifteen temperate-zone vespertilionids for which data were summarized by Gaisler (1979) had a mean adult survivorship of 71%, ranging from 57% to 86%. By contrast Fleming (1975) reported survival rates of <5% to 45% in 24 species of tropical rodents in a comparable size range. The survival rates of adult bats of both temperate and tropical regions are unusually high, and are comparable to those of much larger mammals. The highest mortality affects bats that are less than a year old. For hibernating species, the critical period seems to be before and during the first hibernation. Newly volant bats are at great risk from predators, and must master the art of prey pursuit and capture very quickly. If they do not, and if they do not accumulate enough fat to last through the winter their chance of survival is slight.

Bats are preyed upon chiefly by hawks and owls, by a variety of small mammalian carnivores, and by snakes. Despite a substantial literature cataloging causes of mortality in bats (Gillette and Kimbrough, 1970), predation is rarely observed by naturalists, and because the fecundity of bats is so low, the conventional wisdom has arisen that bats suffer low rates of predation. Tuttle and Stevenson (1982) suggested that predation rates are underestimated by chiroptologists. Still, there is not enough quantitative data to allow valid generalizations.

Fecundity

Most bats have one youngster per litter once or twice a year. All temperate zone bats are monestrous. Monestry is unavoidable in the temperate zone given the usual regimen of hibernation, delayed fertilization, prolonged gestation, infant dependency, and the shortness of the season of food availability. Tropical bats are monestrous or polyestrous and generally produce one infant per birth. A common pattern is seasonal polyestry timed to coincide with the start of rainy periods (Wilson, 1979). *Carollia* has about two young per year (Fleming, 1988), and this seems to be true of most phyllostomids (Humphrey and Bonaccorso, 1979). Tree-roosting bats, such as *Lasiurus*, may have litters of up to five, although litters of two to three are more usual. Mean litter size of the 30 vespertilionids recorded for the United States (Barbour and Davis, 1969) is 1.55, while that of those 37 tropical bats, mostly neotropical phyllostomids, which have been the subjects of accounts in the periodical *Mammalian Species*, is 1.00 ($\chi 2 = 18.3$, $P < 0.001$). As for avian clutch size (e.g., Cody, 1971), litter size in bats seems to increase with latitude.

In eutherian mammals, litter size is negatively correlated with weight. The relationship is given by Calder (1984) as

$$\text{Litter size} = 3.43 \ W^{-0.16}.$$

The average litter size of a 10–20 g bat should be 6.4 to 7.2. Bats are most unusual among small mammals in the low number of young they produce per birth (Fig. 1.2). The mean annual production by female bats has been estimated by Gaisler (1979) at between 0.5 and 1.5 young, lower than that of any vertebrates other than the larger mammals.

Gestation and infant dependency

Gestation and infant dependency last a long time in bats; for their size they are outstanding in this respect, and most closely match the primates (Eisenberg, 1981). Even the smallest bats have gestation periods of about 2 months, and 3 or 4 weeks to several months elapse before the young are completely independent. For most temperate zone species, reproductive maturity does not come before 9 months to a year, and the same may be

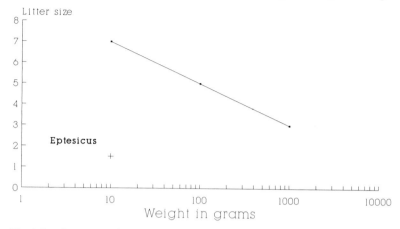

Fig. 1.2. Regression line showing the usual relationship between weight and litter size in small mammals (Calder, 1984). A 10-g bat is plotted for contrast.

true for many tropical kinds. Female *Carollia* normally do not breed until they are 1-year old, although some individuals may mature sexually by 2.5 months (Fleming, 1988). In contrast, few comparably sized rodents have gestations or infant dependency periods much in excess of a month.

Population fluctuations

Fleming's 9-year study of *Carollia* demonstrated relatively constant population densities (Fleming, 1988). Demographic data suggested a net reproductive rate (R_0) of approximately 1.0. A 12-year study of two hibernating populations of *Pipistrellus subflavus* in West Virginia (Davis, 1966) showed one to be relatively stable with a coefficient of variation of population size of 16%, while the other experienced several unusually high years, and had a coefficient of variation of 39%. After a study of Czechoslovakian bat populations, Gaisler (1979) concluded that they were relatively stable. Everything about chiropteran reproductive biology suggests that their populations cannot fluctuate markedly.

Escape in time: heterothermy and dormancy

Perhaps 10% of the known species of bats hibernate. These are animals of the seasonal temperate zones, mostly belonging to the families Vespertilionidae and Rhinolophidae. Most information on hibernation comes from studies carried out in Europe and North America. Typically, hibernating bats become fat in late summer and seek out a hibernaculum, commonly a cave, where large numbers aggregate and enter a dormancy which lasts until spring. During the dormant period most bats arouse periodically. The necessity for hibernation imposes severe constraints on reproductive behavior and physiology. Males undergo spermatogenesis in late summer and store sperm in the epididymides through the winter. Copulation may take place just before hibernation, upon emergence in spring, or sometimes during arousal periods in hibernation. Females mated in the fall store sperm in the uterus, and develop a large ovarian follicle. Upon emergence in the spring, ovulation and fertilization take place, followed by a 1–3-month gestation period. This well-studied pattern is often thought of as typical of chiropteran life, but this reveals the bias of temperate-zone biologists. Only bats living in the temperate-zone are like this. A substantial majority of bats is exclusively or mostly tropical, and most do not seem capable of entering prolonged periods of dormancy. In the face of lowered ambient temperatures many tropical species maintain their body temperatures at normal operating levels. Some types, when experimentally subjected to lowered temperatures for a prolonged period, at first thermoregulate and then succumb to hypothermia and death. Others may allow their body temperatures to decline during the day, or even for several days (Brosset, 1961), but no tropical bats, so far as known, become dormant for extended periods in the manner of temperate-zone species. Nonetheless, a certain amount of thermal lability is present in many tropical bats, and in one view (Hill and Smith, 1984) this facultative heterothermy preadapted them for temperate-zone life, and may have formed the basis for invasion of the temperate zones by vespertilionids and rhinolophids.

In summary, the ability to escape unfavorable conditions through dormancy, and thus, in effect, to occupy a relatively stable environment, is essential to occupancy of temperate zones by small insectivorous bats. It seems not to be a dominant factor in the biology of bats as an order.

Escape in space: migration

Many temperate-zone bats that hibernate make annual trips, sometimes extended ones, to and from suitable hibernacula. While this behavior is technically defined as migration, the essential part of the pattern is the hibernation interval and not the migratory movement per se. Migration of the sort practiced by many birds, moving from a summer breeding ground to a wintering area where the animals remain active, has been demonstrated for only a handful of species. In North America, migration of the avian kind has been demonstrated for the guano bat *Tadarida brasiliensis* (Cockrum, 1969), banded individuals of which have been recovered after flights of over 1200 kilometers. The hoary bat *Lasiurus cinereus* shows a clear latitudinal pattern of seasonal change in distribution (Findley and Jones, 1964), but what the activity of these animals may be in their winter quarters is not well known. Red bats, *Lasiurus borealis*, also migrate southward in the winter, but have been recoverd dormant in trees while on their winter range. Lasiurine bats roost in the foliage of trees, and it may be that many move southward far enough so that hibernation in these exposed roosts is possible. In Europe several genera of bats make long seasonal flights, but in most cases these may simply take them to suitable hibernacula. The distance record seems to be held by a noctule, *Nyctalus noctula*, reported by Krzanowski (1964) to have travelled 2347 kilometers from Russia to Bulgaria.

Some populations of the nectar-feeding phyllostomid bats *Leptonycteris* and *Choeronycteris* appear in the southwestern United States in summer, when cacti, yuccas, and agaves are in bloom, and bear their young in caves and rock shelters. In late fall, they depart for Mexico where they remain active throughout the winter.

Larger species of Old World fruit bats, Pteropidae, may migrate seasonally to take advantage of fruiting seasons of trees. The African *Eidolon helvum*, the range of which is centered in equatorial regions, migrates to the north and as far south as South Africa. The Old World vespertilionid *Miniopterus schreibersi* carries out extensive travels practically throughout the warmer parts of Eurasia and Africa.

Overall, the record of migration by bats is not impressive, and for the order as a whole, it is a minor phenomenon.

Bats compared with other vertebrates

Bats are unique among mammals of their size in their long lives, low fecundity, duration of maternal care, and slow development. How do they compare with other groups of vertebrates which are commonly studied by community ecologists?

Comparably sized birds, especially tropical passerines, are somewhat more like bats ecologically than are small terrestrial mammals. Approximately two broods per year are usual for such species, but clutch size per brood ranges from over 2 to 14 (Cody, 1971), so that annual fecundity is higher than for bats. Survival rates of small birds in the tropics may be close to those of bats: 80–90% has been reported, but temperate-zone passerines typically experience much lower survival, in the 25–65% range (summarized in Cody, 1971). Small birds generally develop more rapidly than do bats, and commonly have incubation periods of approximately 2 weeks, followed by a 2–3-week interval of nestling dependency. Individual birds may live a long time, but the maximum age recorded for a banded wild passeriform, about 20 years (Welty and Baptista, 1988), is exceeded by many chiropteran records, and the fact that many more birds have been banded than bats suggests that bats are comfortably in the lead in longevity. Small birds as a group have demonstrably higher fecundities and shorter lives than bats.

Temperate and tropical lizards vary considerably in their life-history charcteristics (Ballinger, 1983). Generally they have a short prereproductive period with high mortality, reproductive life spans of varying length, clutch sizes from 1 to 70, and single or multiple broods, depending on latitude and seasonality. Some lizards may grow continuously at a decreasing rate, but most are determinate growers. Fecundity increases with body size within groups, and especially within species, but some species show no such relationship. Life spans are quite variable. Geckos (family Gekkonidae), almost all of which are tropical, have only one or two eggs per clutch, but reproduce continuously in the non-seasonal tropics. The 700 gecko species comprise 20% of all lizards. Many neotropical anoles (*Anolis*, family Iguanidae), produce a single egg at a time, perhaps an adaptation to arboreal life. Other species of small iguanids produce litters (or clutches) of 6–20. In the spiny lizards (*Sceloporus*), viviparous species have fewer young (mean 9.6) and are single-brooded, while oviparous kinds have larger broods (mean 12.5) and are multi-brooded. Age at sexual maturity in lizards ranges from 2 months to several years (as high as 15 years in some iguanids) and there seems to be little information on survivorship, although some of the long-lived iguanas show less than 2% annual mortality. The scant evidence suggests that geckos and anoles may approach bats in low fecundity, but demographic data on tropical lizards are not such that good generalizations concerning survival and longevity are possible.

Much interest has focused on fish communities, especially those of coral reefs and tropical streams and lakes (Lowe-McConnell, 1987). Life-history strategies and demographic parameters for fish vary greatly, depending upon environmental stability. Many stream fish live in fluctuating environments where floods may quickly produce huge areas of food-filled habitat. Fish in these situations may reproduce prolifically, only to suffer immense mortality when the floods abate and the aquatic habitat diminishes. In the seasonal tropics, lakes may wax and wane. A few fish have adapted to the unpredictable availability of water by producing drought-resistant eggs, which are dormant until the rains come. Some lungfish survive the drought as torpid adults. Inhabitants of stable tropical swamps and marshes of southeast Asia, such as fighting fish and gouramies (Belontiidae), build nests and lay a few eggs which are guarded by the males. The reproductive lives of reef fish are shrouded in mystery. Spawning takes place frequently but the eggs and larvae of most species are planktonic, essentially disappearing from the purview of fisheries biologists until the transforming young settle on the reef. The timing and frequency of recruitment, and the number and species of recruits may be decoupled from the configuration of the adult community. Whatever else can be said about this system, it is clear that large numbers of eggs and frequent spawnings are necessary to compensate for the uncertainties of larval life in the pelagic realm. Survival in fish as in many other organisms is inversely related to fecundity. Aquarium fish rarely attain an age of 30 years, and a majority of smaller wild species probably live 2 years or less (Lagler, Bardach and Miller, 1962; Moyle and Cech, 1982). As with the other vertebrates we have discussed, few groups of smaller fish approach bats in reproductive and demographic features.

Among vertebrates, it seems, very few, with the exception of rather large species, exhibit the low fecundity and high longevity of bats. Bats are obviously outstanding K-strategists, certainly much more so than other vertebrates that community ecologists like to study. Why should this commend the Chiroptera to the attention of the community biologist?

Bats as small, K-selected vertebrates

In the tropics the environment, in the popular view at least, is relatively stable and predictable, and should favor organisms that can use the resources most efficiently and that can deal effectively with competition from other organisms. Here the problem of responding to harsh, unexpected environmental perturbations is minimal. In many parts of the temperate zone, however, environmental conditions are harsher, less stable, and less predictable. Here the premium should be on ability to respond rapidly to environmental disaster or opportunity. High reproductive capacity should enhance that ability. This familiar idea was first articulated by Dobzhansky (1950) and later formalized by MacArthur (1962) and MacArthur and Wilson (1967) as r- and K-selection. Pianka (1970) expanded on the concept and provided a table listing attributes of r- and K-selected organisms. He noted that the qualities of K-selected organisms (e.g., relatively constant population size, greater competitive ability, slower development, iteroparity, greater longevity, energetic efficiency) are correlates of greater body size. Calder (1984) summarized allometric equations describing the relationship between size and the various attributes of r- and K-selected species. The interest of community ecologists no longer focuses on the r–K concept. Nonetheless the relationship between size and ecological and physiological traits is pervasive, though not universal. The biology of bats, as we have seen, seems a serious contradiction to this generalization.

Based on their size, bats should be relatively r-selected among mammals. They should be short-lived, rapidly reproducing organisms exhibiting pronounced population fluctuations, and their habitats should be characterized by impermanence and unpredictability. All of these predictions are wrong. As we have seen, bats are long-lived, slowly reproducing creatures which maintain relatively stable populations. This contradiction raises an important 'how' and an important 'why' question. The 'how', or proximate, question asks: what ecological and physiological processes conspire to promote unexpectedly long life in such a small mammal? This is a question for autecologists and physiologists. It is dealt with only briefly in this book. The 'why', or ultimate, question asks: what selective factors have favored bats which possess the suite of attributes which characterize K-selected animals? This is a question for evolutionary ecologists. The answer that flows from life history theory is: the demands of life in a stable, predictable environment wherein the chief adaptations required enhance the animal's ability to deal with biotic, rather than physical, interactions. If we want to study communities of K-selected animals, bats provide one of the best opportunities. And, since many small vertebrates which have been studied fall toward the r-end of the spectrum, these investigations should add needed balance to our perspective. So we must question whether bats do indeed exemplify the patterns and processes to be expected of K-selected organisms. Pursuit of that answer is one of the important purposes of this book.

Summary

The study of communities is rewarding because, among other things, it helps us toward the understanding needed to manage them in a prudent way. To the extent that communities are controlled by biotic interactions among organisms, those community studies which illuminate these interactions are of especial interest. Assemblages of closely related species may be most likely to provide examples of these interactions, and the relationships of interest are most likely to be well developed among K-selected forms occupying stable environments. In this sense, bats seem to be ideal model organisms.

2 · *An introduction to bats*

There are about 4200 living species of mammals, and almost 1000 of them belong to the order Chiroptera. Only rodents, with over 1700 species, are more speciose. Most species of bats, about 88%, are exclusively tropical. Most of the 12% which live in the temperate regions are members of the family Vespertilionidae, (96 of 330 vespertilionids are temperate zone species). Most of these belong to the widespread genera *Myotis* (40 of 94 species are temperate), *Pipistrellus* (11 of 48 species), and *Eptesicus* (9 of 33 species).

Table 2.1 and Figs. 2.1 and 2.2 summarize some quantitative, geographic, and ecological facts about bat families. The following brief descriptions provide an introductory guide to the diversity of bat groups. The classification is mostly based on that used in Hill and Smith (1984).

Order: Chiroptera. Bats first appeared in the fossil record of Eocene tropical forests as fully developed volant animals, much like those living today. They presumably arose from small arboreal insectivorous mammals, and the concensus seems to be that they began as gliders, perhaps somewhat like modern colugos, tropical gliding mammals of the order Dermoptera. Two very distinct suborders of bats are recognized.

Suborder: Megachiroptera. These are the medium to large, mostly frugivorous bats of the Old World tropics, referred to as flying foxes, fruit bats, or simply as megabats. All belong to the family Pteropidae. Megabats are so different from microbats, members of the suborder Microchiroptera, that some chiroptologists deny that the two suborders arose from a common ancestor. Unlike microbats, megabats do not use high-frequency echolocation, although a couple of species use a distinctive, and probably independently evolved, low-frequency kind based on tongue-clicking. Megabats have large eyes and well-developed vision. The wings and flight anatomy are much less sophisticated than in

Table 2.1 *A survey of bat families*

Family	No. genera	No. species	Distribution	Trophic/foraging
Megachiroptera				
Pteropidae	42	175	Paleotropical	Frugivores
Microchiroptera				
Emballonuroidea				
Rhinopomatidae	1	3	Old World. deserts	Aerial insectivores
Emballonuridae	13	51	Pantropical	Aerial insectivores
Craseonycteridae	1	1	Thailand	Aerial insectivores
Rhinolophoidea				
Rhinolophidae	10	129	Old World	Insectivores, carnivores,
Nycteridae	1	12	Paleotropical	gleaners, 'fly-catchers',
Megadermatidae	4	5	Paleotropical	aerial pursuers
Phyllostomoidea				
Phyllostomidae				
Phyllostominae	12	35	Neotropical	Omnivores, gleaners
Glossophaginae	13	33	Neotropical	Nectarivores, frugivores
Carolliinae	2	7	Neotropical	Frugivores
Stenodermatinae	18	62	Neotropical	Frugivores
Brachyphyllinae	3	7	Antillean	Frugivores, nectarivores
Desmodontinae	3	3	Neotropical	Sanguinivores
Mormoopidae	2	8	Neotropical	Aerial insectivores
Noctilionidae	2	2	Neotropical	Piscivores, insectivores
Vespertilionoidea				
Natalidae	1	4	Neotropical	Aerial insectivores, gleaners
Furipteridae	2	2	Neotropical	Aerial insectivores, gleaners
Thyropteridae	1	2	Neotropical	Aerial insectivores, gleaners
Myzopodidae	1	1	Madagascar	Aerial insectivores, gleaners
Vespertilionidae	41	330	Cosmopolitan	Aerial insectivores, gleaners
Mystacinidae	1	2	New Zealand	Omnivores, terrestrial
Molossidae	13	89	Pantropical	Aerial insectivores

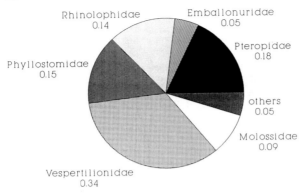

Fig. 2.1. Relative world-wide abundance of the commonest bat families.

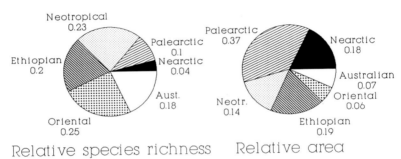

Fig. 2.2. Relative species richness and land areas of the classical zoogeographic regions.

microbats, and details of dental and cranial anatomy suggest remote relationship at best. Megabats feed upon fruit, flowers, or pollen. Rather than biting off and swallowing mouthfuls of fruit, many kinds seem to crush the fruit, swallow the juices, and spit out the pulp. Several small species have elongate snouts and tongues and specialize in nectar and pollen feeding. Most megabats roost in trees, sometimes aggregating in huge flocks. Several echolocating species (*Rousettus*) have access to caves as roosts.

Suborder: Microchiroptera. The perceptual world of microbats is built upon high-frequency sound. Sounds in the 20–150 kilohertz range are produced in the larynx, and the returning echoes guide the bats in their pursuit of prey and avoidance of obstacles. Much of the sometimes bizarre external facial and aural anatomy represents adaptation to the emitting, focusing, and receiving of ultrasound. The flight of microbats is more maneuverable than that of megabats, and the anatomy of the wings and shoulders more complicated. Microbats at roost fold their wings and extend their heads dorsally, while megabats wrap themselves in their wings and look ventrally, and the structure of the cervical vertebrae is modified accordingly. The 16 microchiropteran families are grouped into superfamilies indicative of presumed phylogenetic relationships.

Superfamily: Emballonuroidea. This group includes the mouse-tailed bats (Rhinopomatidae), the sheath-tailed bats (Emballonuridae), and the bumble-bee bat (Craseonycteridae). All are aerial insectivores. The validity of the grouping is somewhat questionable. Rhinopomatids have recently been shown to have affinities also with the Vespertilionidae (Robbins and Sarich, 1988).

Family: Rhinopomatidae. Mouse-tailed bats. The three species are chiefly desert bats of North Africa and Southwest Asia. They are aggregating, insectivorous animals roosting in rock shelters and human structures. They may become fat and enter dormancy in seasons of environmental stress. Their kidneys are highly modified for the production of concentrated urine, and they may be the most specialized of bats for desert life.

Family: Craseonycteridae. Bumble-bee bat. This family comprises one species, *Craseonycteris thonglongyai*. The bumble-bee bat was discovered in west central Thailand in 1973. This is the smallest bat, with a forearm of 22–26 mm, and may be the smallest mammal. It roosts in high-ceilinged limestone caves, and has been observed foraging over bamboo groves. The species is rare, and the region in which it occurs has been extensively deforested since the bat was first discovered.

Family: Emballonuridae. Sheath-tailed bats. These pantropical bats are mostly small-eared aerial insectivores. A number of New World species are polygynous, living in male-controlled harems. Roosting may be in caves or in the open on tree trunks or in the foliage. Roosting individuals are in well-lighted situations, remain alert, and take flight quickly when disturbed. At roost, the tail, which penetrates the dorsal surface of the tail membrane, is carried curled dorsally. Roosting bats are very alert with their forearms spread and heads raised. Males of some New World

Fig. 2.3. The neotropical emballonurid *Rhynchonycteris naso*. These tiny insectivorous bats roost in groups lined up on the stems of small trees, often overhanging forest streams. The bats may forage as a group, following each other in single file over the watercourse. (Photo by Scott Altenbach from Finca La Selva, Heredia Province, Costa Rica.)

species have glandular pockets in the skin of the anterior flight membrane which produce pheromones used in courtship. Most of the species seem to belong to a monophyletic group of New World origin (Fig. 2.3).

Superfamily: Rhinolophoidea. A palaeotropical autochthonous unit, this group includes the Old World leaf-nosed bats or horseshoe bats (Rhinolophidae), the false vampires (Megadermatidae), and the slit-faced bats (Nycteridae). All emit their ultrasound nasally and possess unusual facial structure, often including complicated protruberances and folds. All are insectivorous or carnivorous. Most are slow maneuverable fliers, and may forage by hovering and gleaning, by hawking from observation perches, or by aerial pursuit. Many are known to orient toward prey-generated sound, and have been referred to as 'flutter detectors.' Rhinolophoid bats represent the extreme development among the Chiroptera of forwardly directed echolocation capability. Frequently the development of nasal foliage is such that it precludes forward vision.

Family: Rhinolophidae. Old World leaf-nosed bats. In these bats, and their close relatives, the Megadermatidae and the Nycteridae, the complexity of facial ornamentation is the extreme seen in the Chiroptera. The nasal excrescences seem related to the production of a narrowly focused beam of sound which is emitted through the narial openings. One subgroup of rhinolophids is sometimes separated as the family Hipposideridae. Rhinolophids are widespread and abundant palaeotropical insectivores. Many roost in caves in large colonies, and a few species become torpid from time to time. About 15 species of the genus *Rhinolophus* have capitalized on this heterothermic ability to colonize temperate Europe and Asia.

Family: Nycteridae. Slit-faced bats. Twelve species in the single genus *Nycteris* occupy tropical forests in Africa and arid regions in the Near East. These broad-winged, large-eared bats capture some of their invertebrate prey by picking it from various substrates, including the ground. Remains of scorpions, non-volant insect larvae, and small vertebrates have been recovered from their stomachs or from beneath their feeding sites. They roost in small groups in caves, in hollow trees or logs, in animal burrows, or in human habitations. Most of the difference between the species is in size, and several species, each of a different size, may be captured in the same location.

Family: Megadermatidae. False vampire bats. The five species of paleotropical false vampires are true carnivores, feeding on such small vertebrates as frogs, bats, rodents, and fish, as well as large invertebrates. Some species hunt by roosting at an observation perch, scanning the area sonically, and then flying out to intercept passing prey. Roosting is in trees, caves and rock shelters, or in the foliage. The African *Lavia frons* is frequently seen flying about in daylight.

Superfamily: Phyllostomoidea. New World leaf-nosed bats (Phyllostomidae), ghost bats (Mormoopidae), and fishing bats (Noctilionidae) are closely related and are considered to have originated and differentiated in the Neotropical region. They are quite diverse in all aspects of their trophic biology and morphology.

Family: Phyllostomidae. New World leaf-nosed bats. All members of this exclusively neotropical family except the vampires have a vertical leaf-shaped flap of skin above the narial aperatures. This nose-leaf is never as complex in its construction as is that of the Rhinolophidae. The ears are large or small, but simply built. Wings, tails, and tail membranes are quite variable. Phyllostomids emit ultrasound through either the narial openings or the mouth. Some kinds rely on nasal sound for orientation. Much of the sound produced by phyllostomids is of very low intensity, and they have been referred to by students of acoustics as whispering bats. Some of their sound, perhaps that used in communication, is of higher intensity (Gould, 1977). The bats in this family pursue a diversity of ways of living. These different modes have resulted in differing physical modifications, which has led to the grouping of the various species in several subfamilies.

Subfamily: Phyllostominae. Most of these bats are generalists in their feeding, consuming a variety of invertebrates as well as fruit in season. A few are more specialized in diet. *Vampyrum* (Fig. 2.4) preys upon birds, and perhaps other vertebrates. *Chrotopterus* captures other bats. *Trachops* (Fig. 2.5) consumes small frogs which it locates by their vocalizations. The most specialized hovering-gleaners among phyllostomids are found in this subfamily, exemplified by *Lonchorhina* and *Mimon* (Fig. 2.6) with their large, forwardly directed ears and extremely elongate nose-leaf. *Macrophyllum* (Fig. 2.7) has enlarged hind feet, like bats in other families which are known to feed upon aquatic prey, and is usually netted over water. As with carnivores generally, phyllostomines are less common than other leaf-nosed species. Their teeth more closely resemble the primitive eutherian tuberculosectorial pattern than do those of other phyllostomids, and their tooth-rows and muzzles are neither shortened nor especially elongate. Roosting is in hollow trees, logs, or in rock shelters.

Subfamily: Stenodermatinae. These are the specialized fruit-eaters among phyllostomids. Their muzzles and tooth-rows tend to be shortened, and their teeth are broad and flattened, with an exterior cutting edge. They range from rather small (10 g) to large (80 g) species, and several different sizes are commonly captured at the same site. Many

Fig. 2.4. The phyllostomine *Vampyrum spectrum*, the largest New World bat. A rare carnivorous species, it feeds upon birds, and sometimes small mammals. The extremely broad flight membranes provide lift for carrying large prey items. (Photo by Scott Altenbach from Finca La Selva, Heredia Province, Costa Rica.)

Fig. 2.5. The phyllostomine frog eater *Trachops cirrhosus*. The facial configuration is typical of hovering-gleaning species, with large, forwardly directed ears. The highly cambered, broad wings provide lift at low air speeds. Although very short-tailed, the tail membrane of this and similar species is held taught by the calcar, seen extending medially from the ankle. (Photo by Scott Altenbach from Finca La Selva, Heredia Province, Costa Rica.)

Fig. 2.6. The phyllostomine *Mimon crenulatum*. The very long nose-leaf and distinctive dorsal white stripe are visible. This species feeds on small arthropods and lizards. (Photo by Scott Altenbach from Finca La Selva, Heredia Provice, Costa Rica.)

species roost in foliage, and white markings, presumably for cryptic purposes, are common. One species, *Ectophylla alba*, is completely white (Fig. 2.8). Within the subfamily, various levels of feeding specialization are seen. Some kinds, for example several *Artibeus*, gather whole fruits which are carried to a feeding roost and consumed. Others, with shorter faces and more reduced tooth-rows, eat holes in larger fruits, and may consume mostly the pulp and juices. Included in this group is the bizarre naked-faced bat *Centurio*, whose lack of hair on the face and head may be a specialization analogous to that of the naked-headed Old and New World vultures which must plunge their heads into the moist interiors of the carcasses upon which they feed. Stenodermatines and Carolliines are usually the most common bats netted in neotropical forests.

Subfamily: Glossophaginae. Flower bats are notable for their elongate rostra, long, protrusible tongues, and reduced dentition (Fig. 2.9). Their

Fig. 2.7. The phyllostomine *Macrophyllum macrophyllum*. These small bats use their long legs and enlarged hind feet to scoop insects, including water striders, from the surface of forest streams and ponds. (Photo by Scott Altenbach from Finca La Selva, Heredia Province, Costa Rica.)

chief specialization is for extracting food from flowers, from which they withdraw nectar and pollen. Like hummingbirds, they have the ability to hover in place while they feed. Other foods are consumed as well, and some of the more generalized species eat fruit and insects. Many roost in caves and rock shelters. The possibility exists that the species assigned to this subfamily are not monophyletic, but represent two separate lineages, each of which has specialized for flower feeding independently.

Subfamily: Carolliinae. The few species of *Carollia* seem to concentrate their feeding on the fruits of pepper plants in the genus *Piper* (Fleming, 1988). These erect, elongate fruits are snapped off by the flying bats and carried to a feeding perch for consumption. Carollias roost in caves and hollow trees in rather large aggregations, of which male-

Fig. 2.8. The stenodermatine white bat *Ectophylla alba*. These tiny frugivores roost in groups under the leaves of heliconias which they alter to make into a tent-like shelter. The proportions, including a very limited tail membrane, no tail, and relatively short, laterally directed ears, and short muzzle, are typical of all stenodermatines. (Photo by Scott Altenbach from Finca La Selva, Heredia Provice, Costa Rica.)

dominated harems are often a feature. Physically, these bats are less specialized than stenodermines, but are more so than phyllostomines. In some places they are the most abundant bats encountered by the field naturalist (Fig. 2.10).

Subfamily: Desmodontinae. In some respects the vampires represent a culmination of the stenodermine trend toward short faces and fruit-gouging, except that the vampires gouge small depressions in the skins of animals, usually endotherms, from which they lap the blood. In this group, the nose-leaf is reduced to a low ridge over the nostrils. The common vampire *Desmodus* feeds chiefly on mammals, often upon domestic livestock. The other two taxa, *Diphylla* (Fig. 2.11) and *Diae-mus*, seem to be bird specialists. Roosting is in caves, rock shelters, and hollow trees. In places where cattle are raised, common vampires may be abundant.

Fig. 2.9. The glossophagine flower bat *Leptonycteris sanborni*. Short ears, elongate muzzles, and reduced nose-leaves are characteristic of this group, most of which visit flowers to feed upon nectar and pollen. Glossophagines play an important role in pollination of various subtropical and tropical plants. (Photo by Scott Altenbach from southern Arizona.)

Subfamily: Brachyphyllinae. These bats, confined to the Greater and Lesser Antilles, are somewhat intermediate between stenodermatines and glossophagines in their specializations and way of life. Their muzzles are elongate, their teeth are somewhat reduced from the typical fruit-eating configuration, and their tongues are extensible. They feed upon nectar, pollen and fruit. Most roost in caves.

Family: Mormoopidae. The moustached bats are so-called because of the elongate hairs surrounding the mouth. In flight, the open mouth and

Fig. 2.10. The carolliine short-tailed fruit bat *Carollia*, probably *C. castanea*. Bats of this genus are among the commonest neotropical bats. Most species feed heavily on the infloresences of plants of the genus *Piper*. Note the incomplete tail membrane, partially supported by the calcars extending from the ankles. (Photo by Scott Altenbach from Finca La Selva, Heredia Province, Costa Rica.)

hairs, together with a series of flaps on the lower lip, form a funnel-shaped opening which may serve to focus the orally emitted echolocation sounds. These long-winged bats are aerial insectivores, capturing their prey on the wing, usually while flying at a relatively low elevation. They may be quite abundant, and roost in large aggregations in hot, humid caves. There are also records of foliage-roosting. The wing membranes attach to the body high on the back, giving the impression in some species that the back is hairless, suggesting the name 'naked-backed bat.' Various lines of evidence indicate that mormoopids are related to

Fig. 2.11. The hairy-legged vampire *Diphylla ecaudata*. Note the short ears, shortened muzzle, and lack of a nose-leaf, characteristics of desmodontines. The pointed, forwardly projecting incisors are also typical. (Photo by Scott Altenbach from Honduras.)

the phyllostomatids, and they were once included with the leaf-nosed bats in the family Phyllostomidae.

Family: Noctilionidae. The fishing bats resemble large mormoopids, but lack the moustache, and instead possess large pendulous upper lips which have suggested the name 'bulldog bat.' Both species forage over water, and the larger, *Noctilio leporinus*, captures small fish with its enormously enlarged hind feet and hind claws (Fig. 2.12). The smaller species, *N. albiventris*, picks insects from the water surface. Both species roost in caves, rock shelters, and tree hollows.

Superfamily: Vespertilionoidea. Included here is a majority of bat species. Vesper bats (Vespertilionidae), free-tailed bats (Molossidae), New World disk-winged bats (Thyropteridae), funnel-eared bats (Natalidae), smoky bats (Furipteridae), and Old World disk-winged bats (Myzopodidae) are all small insectivorous species, greatly resembling one another in morphology. New Zealand short-tailed bats (Mystacinidae) are included here, but recent studies suggest that they may be more closely allied to the phyllostomoid complex.

Fig. 2.12.   The fishing bat *Noctilio leporinus*. The enlarged toes are dragged
through the water in an attempt to gaff a piece of food. Note the long legs and
calcar which helps to hoist the tail membrane above the water. The short tail may
be seen protruding posteriorly. Very long and broad wings provide lift at low
speeds as the bat approaches the water surface. The large pendulous upper lips of
this species have given it the name 'bulldog bat.' (Photo by Scott Altenbach from
western Mexico.)

Family: Natalidae. The funnel-eared bats are very small with long slender limbs, broad flight membranes, forwardly directed, funnel-shaped ears, and unusually small eyes, which are so placed that forward vision appears impossible. Their flight is very slow and highly maneuverable, suggesting that they forage by hovering and gleaning. They roost in large colonies in hot humid caves. Naturalists who have handled these bats note that they seem excessively subject to dehydration. Even in living individuals, the flight membranes begin to dry if the animals are kept for long in a dry environment. Natalids are confined to the neotropics, and seem closely related to two other small neotropical families, the Thyropteridae and Furipteridae.

Family: Furipteridae. The two species of smoky bat somewhat resemble natalids, and like them, are limited to the neotropics. Their thumbs are tiny and almost hidden in the flight membrane. Roosting is in caves or hollow logs. Nothing is known of the lives of these animals.

Family: Thyropteridae. The neotropical disk-winged bats are like small natalids with tiny suction cups on their wrists and hind feet. These devices are used in adhering to the smooth surfaces of leaves. The bats are known to roost in the rolled leaves of *Heliconia*, a relative of banana common in neotropical forests. The small roosting groups stay together night after night, frequently changing their roosting site as the leaf expands and becomes unsuitable as a shelter.

Family: Myzopodidae. The one species of Old World disk-winged bat is confined to Madagascar. These animals are provided with wrist and foot disks as are thyropterids, although the details of construction are different. The few observations suggest that rolled leaves and the axils of traveller's palms, *Ravenala*, are used as roost sites (Randolph Peterson, personal communication).

Family: Vespertilionidae. This is the largest of the bat families, and the second largest family of mammals. The common name 'vesper bat' or 'evening bat' appears in books, but is rarely used since many of the individual genera have widely known common names. As a family, vesper bats lack prominent specialization. They are mostly oral emitters of ultrasound and lack facial excrescences. Their ears may be large or small, but are mostly without peculiar configurations. All have tails which extend to the end of the rather extensive tail membrane. A

Fig. 2.13. The hovering-gleaning vespertilionid *Plecotus townsendii* in slow flight. Enormous, forwardly directed ears and broad flight membranes are a part of the morphological repertoire of this and similar species. The tail extends to the end of an extensive tail membrane, used in insect capture in this and many vespertilionids. (Photo by Scott Altenbach from New Mexico.)

majority are gray, brown, or blackish in color. Vespertilionids occupy most habitats available to bats, and are world-wide in distribution, excepting only Antarctica, Greenland, the extreme northern parts of North America and Eurasia, and the Arctic archipelago. Most are aerial pursuers of insects. Some, such as *Plecotus* (Fig. 2.13), practice hovering-gleaning. Various kinds of *Myotis* engage in water-surface foraging. One or two are known to capture prey on the ground, for example *Antrozous* (Fig. 2.14). Caves, rock shelters, hollows of various kinds, buildings, and foliage may be utilized as roosts. Many, perhaps most, are heterother-mic, and in the temperate zones enter hibernation in the winter. A few

Fig. 2.14. The desert pallid bat *Antrozous pallidus*. Relatively shorter ears and larger eyes than those seen in *Plecotus* suggest the importance of vision in orientation. This species often lands on the ground to pursue arthropods and small vertebrates. It has one of the most specialized kidneys among southwestern American desert bats. The full tail membrane and long tail are typical of vespertilionids. (Photo by Scott Altenbach from New Mexico.)

species may migrate considerable distances. Over half of the known species belong to three genera: *Myotis*, *Pipistrellus*, and *Eptesicus*. *Myotis* and *Eptesicus* are cosmopolitan, and *Pipistrellus* is absent only from South America. Vespertilionids occur throughout the tropics, and are almost the only bats found in the north temperate zone.

Family: Mystacinidae.   The one or two species of short-tailed bats are confined to New Zealand. The animals are distinctive among bats in the degree to which they are specialized for terrestrial locomotion by means of special pockets for the furling of delicate parts of the flight membranes. Food is captured on the ground as well as from trees, and fruit, nectar, and pollen, as well as insects, are consumed. Roosting is in hollow trees. Brief hibernation takes place in winter. The phylogenetic relationships of *Mystacinus* are not clear, but recent speculation allies it with the phyllostomoids.

Family: Molossidae.   Free-tailed bats as a group are specialized for rapid aerial pursuit of insects and probably for seeking widely scattered insect concentrations over considerable distances. Long narrow wings and retractible tail membranes reduce drag and enhance speed, and laterally placed ears and eyes provide a wide lateral field of perception. Some kinds, such as *Eumops perotis* and various species of *Tadarida* and *Nyctinomops* (Fig. 2.15), forage many miles over arid regions and may specialize in locating nuptial flights of termites. In tropical regions, the molossid domain is the airspace above the forest canopy, where their swift-like flight is distinctive. Ultrasound of high intensity is orally emitted, and may enter the sonic range. In some regions the intense audible cries of molossids are a familiar sound to the chiroptologist. Most molossids are also adapted to locomotion in crevices and other confined spaces. There they crawl rapidly forward and backward, and some species have a markedly flattened cranium which allows access to the narrowest slits. Roosting is in caves, rock crevices, or sometimes in hollow trees and buildings. Most species appear to be insectivorous. The famous Brazilian free-tail or guano bat, *Tadarida brasiliensis*, undertakes substantial migrations.

Summary

Most of the nearly 1000 species of bats are tropical. The most important family in the temperate zone is the Vespertilionidae. In the Old World

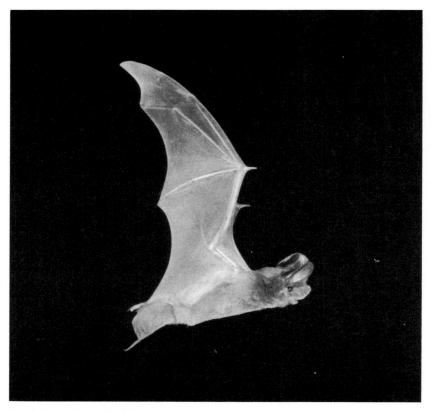

Fig. 2.15. The molossid *Nyctinomops macrotis.* Long, narrow wings, and a long tail, which extends beyond the tail membrane, typify molossid bats, most of which are high-speed and long-distance aerial foragers for insects. The tail membrane can be retracted during high-speed flight to reduce drag. It has been suggested that the long, forwardly and downwardly directed ears of *Nyctinomops* and the related *Eumops* and *Otonycteris* function as auxiliary airfoils to help support the head in flight. It has been suggested, alternatively, that the orientation and size of the ears may help the animals in detecting the echos of their low-frequency cries as they return from the ground. (Photo by Scott Altenbach from New Mexico.)

tropics, the Pteropidae are the dominant frugivores while rhinolophoids, vespertilionids, and molossids are the chief insectivores. The New World tropics are dominated by the Phyllostomoidea, which occupy frugivorous, nectarivorous, carnivorous, piscivorous, insectivorous and sanguinivorous niches, while emballonurids and molossids are the principal aerial insectivores. Pteropids and rhinolophoids are palaeotropical autochthons, while the phyllostomoids, thyropterids, furipterids, and natalids are the autochthonous neotropical units. One section of the Emballonuridae appears to be autochthonous in the New World as well.

3 · *Methodology in the study of bats*

The suitability of a group of organisms for ecological study depends importantly upon the ease of observation and data gathering. Birds have been the favored larger animals for ecology because it is easy to see and identify them, and their behavior is relatively easy to observe and document. For a group of animals to be useful to the ecologist it must be possible to find out where they are, what kinds there are, how many there are, how individuals and species interact, what they eat, what eats them, and where they seek shelter. How do bats rank in these categories? I think it is fair to state that successful bat ecologists are rarely conscripted from the ranks of dilettantes!

A recent summary of methodology in ecological and behavioral study of bats has been provided by Kunz (1988), and many of the remarks that follow are based upon that source.

What bats are there?

To answer this question it is usually necessary to capture the bats. Those bats that can be observed visually at dusk or after dark can rarely be identified with certainty. Bats are best identified in the hand. Even in the hand, many cannot be accurately determined casually, especially in the light of a flashlight or lantern. Often it is necessary to retain the animals overnight so that they may be examined by daylight in the laboratory where identification aids are available. In tropical regions, certain identification frequently cannot be accomplished without sacrificing the bat and examining its cleaned skull. The most experienced chiropteran systematists hesitate to make field identifications of all living specimens except in very well-studied areas. Identification keys are increasingly available for most parts of the world, but the vocabulary needed to use them, that of systematic mammalogy, must be acquired. Even with a preserved specimen and a good key, certainty cannot always be attained without comparison with specimens of known identity in a good museum. Usually one cannot accurately identify bats without a certain amount of discipline and rigorous preparation.

The commonest devices for capturing flying bats are mist nets, structures of braided or monofilament nylon with a usual mesh size of 36 mm and an overall size of 2 m by 6 to 36 m. These nets are set across water holes, streams or trails, or in a variety of other places. Mist nets were first designed to capture birds for the commercial market in Europe and the Orient. Later, scientific bird collectors began to use them, and in the tropics found that if the nets were left in place overnight they captured bats (Van Tyne, 1933). By the mid 1950s, chiroptologists pressed them into service. This date is fairly important to keep in mind when assessing bat literature: little quantitative data on free-flying bats are available from before this time, just as reef-fish biology was largely limited to systematics before the advent of scuba. Mist nets are usually set at ground level, and so capture a selected subset of the local bat fauna, usually failing to take species that forage in or above the canopy. Nets are especially effective at capturing frugivorous species, and insectivores are typically underrepresented. When set across forest trails and streams, the take from a net represents mostly animals going to or coming from foraging areas, and nets over other bodies of water sample foraging or watering bats. Thus there are a number of biases in data gathered from mist-netting, and a balanced picture of the local bat community is rarely obtained. Nets must be watched continuously. If left unattended, the captured bats in their struggles may become so entangled that they can be extracted only with difficulty, and perhaps with injury to the bat. In addition, predators often attack bats in unguarded nets. Tuttle, or harp, traps consist of metal frames with fine wire stretched vertically across the frame at close intervals. Bats striking the wires slide down into a bag where they may hide until the collector removes them. Although less frightening for the bats, the Tuttle trap has many of the other biases of the mist net, except that it is more effective at capturing small insectivorous species.

Before the advent of mist nets, the commonest method for sampling bats was by locating their roosts. Many bats roost communally in caves, rock shelters, hollow trees, or buildings, and a great deal of information on kinds inhabiting a local area can be gathered by visiting such a place. Identification without collection is difficult for many species, however, and disturbance of bats in their roosts often results in the animals abandoning the site, an especially serious blow to the animals, since suitable roosts may be in short supply. In the tropics many species roost in foliage, in the open on tree trunks, or in small retreats under roots, cut banks, or even in rodent burrows and spider nests. Locating bats in such places is probably not possible in any systematic way, especially if the foliage roosts are high in the canopy of the forest.

It has become increasingly possible to identify flying bats by means of their sonic signatures on a field oscilloscope after careful study of the sonograms of identified species in the laboratory. Fenton and Bell (1981) studied vocalization of the sympatric insectivorous bats in Ontario (2 species), southern Arizona (10 species) and in Zimbabwe (23 species). At the North American stations, accurate identification was possible using the visual display of the calls produced by a period meter but, in Zimbabwe, the large number of species made accurate identification possible only after study of recordings. Crome and Richards (1988) used this method succesfully in the rain forests of Queensland to study the composition of insectivorous bat communities foraging in forest gaps and in the canopy. They were able to resolve the cries of 12 species, but noted that several other kinds that are of possible occurrence in their study area have such low-intensity cries that the method would not have detected them. This approach would not be useful for megachiropterans, which do not echolocate, or for the large number of whispering phyllostomids in the neotropics. At Finca La Selva, Costa Rica, a broad band detector rarely records the passage of a bat in the deep forest where hundreds of phyllostomids may be captured in mist nets set across trails. However the same detector, pointed toward the sky in a clearing or at the forest edge along a river, records a cacophony of foraging insectivorous bats (personal observation).

In summary, the methods currently available to detect and identify bats are time-consuming and require a certain level of systematic, and perhaps electronic, expertise. Moreover all the methods have strong biases in the kinds of bats detectable and in the habitats from which information comes. The ecologist who wishes to gain a balanced picture of the bat community at a given station must spend a substantial amount of time and utilize a variety of methods before an acceptable level of confidence in the results can be achieved.

How many bats are there?

Methods for estimating the population densities of bats all suffer from a critical weakness: the size of the area from which the sample is being drawn is never known. Thus reasonable estimates of bat density per unit area are almost non-existent. Nonetheless, various estimation methods do make it possible for regions and habitats to be compared in a relative way.

Marking, releasing, and recapturing bats, and then calculating abundance on the basis of the proportion of recaptured bats that have been marked using the Jolly–Seber (Seber, 1982) or a similar method is an approach open to bat netters. The assumptions of the method, however, are often likely to be violated for bat collections. One such assumption is that marked individuals are as likely to be retaken as unmarked ones. In some studies (Stevenson and Tuttle, 1981), marking has clearly been shown to have a long-term affect on the activity of the bat. Moreover, conventional wisdom among bat collectors suggests that previously netted bats are more difficult to recapture. Bats that have been marked with wing bands often suffer injury, and perhaps some debility in flying, as well as increased mortality. Other methods, such as marking wing membranes with patterns of small punctures, may not carry these disadvantages. Marked animals are assumed to mix randomly with the study population. For most bats this is almost certainly not valid. Many bats are associated with social groupings, perhaps families or harems, and upon release return to their group, rather than wandering randomly through the conspecific population. In other cases, especially among temperate zone vespertilionids, males and females segregate in the breeding season, so that within a given species two subpopulations exist, perhaps each with different likelihoods of entering traps.

Zippin (1956) and Southwood (1966) described the removal trapping method of population estimation, a technique potentially useful for bat and bird ecologists. The technique seemingly was first used for bats by Fischer (1973) and for birds by MacArthur and MacArthur (1974). A least–squares regression is computed of rate of capture per hour of new individuals against cumulative number of individuals taken. The x-intercept is taken as the population estimate. Fischer discussed uses and limitations of the method for bats.

The number of bats in roosts has been estimated in several ways. In large roosts in caves where bats are closely packed, the number occupying a known area may be counted, and then the total area occupied estimated. More precisely, photographs of the roosting clusters may be taken and the animals counted on an enlargement. Numbers of bats entering or leaving roosts have been determined by photographic or acoustical methods. In a few cases it has been possible to locate tree-roosting species and estimate their numbers in limited areas (Constantine, 1966). The neotropical disk-winged bat *Thyroptera tricolor* roosts in the rolled young leaves of *Heliconia* plants, and in one study most of the bats in a limited forest area in Costa Rica where heliconias grew were located. Thus an absolute calculation of density for that specific plant community was possible (Findley and Wilson, 1974). Counts of roosting aggregations suffer from the limitation inherent in all bat census methods, that the area utilized by the roosting aggregation is not known. However studies of the numbers and distribution of bats in roosts provide important data for community ecology. It is likely that in many situations roosts are factors limiting the occurrence of bats. Ways in which different kinds of bats share roosts may be as important to an understanding of the structure of bat communities as knowledge of the ways in which food resources are allocated.

In a few rare instances it has been possible to locate visually most of the hunting beats for an insectivorous bat species, and to determine the absolute number of bats per unit area. Nyholm (1965) was able to accomplish this for *Myotis mystacinus* in Finland, where the long summer twilight allowed him to observe the bats foraging, each in its own area, in forest clearings.

Determining the absolute number of bats of different kinds inhabiting or foraging within a given area is for the most part beyond the reach of readily available technology at the moment. Perhaps the best that can be done by the community ecologist is to assess the relative abundance of different species and to compare regions and habitats with respect to the numbers of bats obtained for given amounts of effort applied to standardized census methods. For example, many naturalists have compared study areas on the basis of bats taken per net per night (the net-night) or per net per hour, or in some cases per meter of net per hour (Findley and Wilson, 1983).

Where do bats go?

To understand how an animal uses its space, it is necessary to know where it goes and how long it spends in each place it visits. Careful observation of time spent in foraging in different parts of trees helped MacArthur (1958) to an understanding of how sympatric wood warblers shared resources in the northeastern United States. Such direct observations of bats are not possible on a predictable basis. Radio-tracking using miniaturized transmitters has made possible substantial advances in our ability to keep track of foraging bats (Barclay and Bell, 1988). Transmitters weighing less than 1 g make it possible to work with bats as small as 15 g. Less expensive are chemoluminescent tags which are visible at a distance of up to 200 m. Teams of observers are needed to follow a light-marked bat over greater distances, but much information can be gathered from an animal foraging in a restricted area. Even cheaper and easier are reflective markers made of Scotch Lite tape. Here the bat must be illuminated with a lantern or flashlight which, if continued, may interfere with normal activity. Long distance travel by bats has been studied most often using samples of banded individuals. Plastic or metal bands are affixed to the forearm, and the hope is that the bat will be recovered in the future. The use of metal bird bands has resulted in a certain amount of damage to the bats, and in large roosts the disturbance caused by capturing and banding large numbers of individuals has had such adverse affects that many workers have given up large-scale banding. Nonetheless, carefully applied bands, used in limited, specifically focused studies, provide one of the cheapest and easiest methods of gathering data on the movements of bats.

How do bats interact?

Most direct observation on the interactions of non-captive bats has been made in roosts. The use of night-vision scopes has enabled the details of some social interactions of cave-roosting phyllostomids to be recorded and a combination of radio-tracking and night vision scopes has been employed to observe group foraging, for example in *Phyllostomus discolor* (Wilkinson, 1987). Closed circuit television and infra-red light were employed by Porter (1978) to observe social interactions in captive *Carollia perspicillata*.

Indirect methods, such as analysis of allozymic polymorphism and karyology, have been used to infer relatedness of members of social groups, and also the paternity of the young in harems (McCracken, 1987). Most of such evidence has suggested that inbreeding is not a common feature of bat population genetics.

Direct observations on how bats of different species interact in their foraging grounds are practically non-existent. Thus ideas about how coexisting bats share space, food, and other resources are mostly inferential.

What do bats eat?

Some information on the diets of bats can be derived from field observations. For example, netted bats often show evidence of their feeding activities through remains of insects in their mouths, or because they are carrying food items. In the neotropics, it is common to net bats, such as *Artibeus*, carrying small fruits, and often the head is dusted with pollen of flowers that have been visited. Frugivorous bats often defecate when netted, and, because of the rapid processing of ingested fruit, the identity of seeds from a recent meal may be obvious.

Many kinds of bats retire to feeding roosts to consume prey or fruits. The remains of food items accumulate under such roosts and provide opportunity for the examination of relatively intact material.

It is often possible to observe bats feeding in and around fruiting and flowering trees, and sometimes the identity of the bat and of the food item ingested or carried away may be determined. Direct observations may also be made under special circumstances of the food taken by insectivorous species. Swarming alates of ants or termites attract many insectivorous bats, and the identity of bat and prey may be matched. *Lasiurus* clips the wings of moths captured in flight, and an alert observer can catch the wings and identify the moth to the specific level (e.g., Merriman, 1990).

Quantitatively, most information that has been reported on bat foods has come from analyses of fecal pellets and stomach contents (Whitaker, 1988). Identification of remains from these two sources is relatively straightforward, but requires a reference collection of likely food items and a good deal of experience and patience. Some biases are attendant upon this method. Probably certain insect groups, such as moths, have a tendency to be overrepresented because the scales from a single ingested moth may appear in fecal pellets produced from several subsequent meals. Soft-bodied forms, such as mayflies, may be underrepresented because few hard chitinous parts are present. Sacrificing bats to examine stomach contents is unwarranted, but the stomachs of specimens captured for other purposes should routinely be saved. Analysis of fecal pellet contents is an easy and non-destructive way to gather a great deal of data. Bats that are taken in mist nets usually defecate within a short time of capture, and may thus be held until pellets can be collected.

What eats bats?

Most information concerning predation on bats is anecdotal, and there seem to be no organized quantitative studies. An encyclopedic compendium of recorded predators has been provided by Gillette and Kimbrough (1970). Most observations have been made as bats enter or leave roosts, when they are subject to attack by hawks, owls, and even snakes. Owls seemingly take bats on a regular basis as attested by the frequent recovery of bat remains from owl pellets, but no systematic technique for study of bat predation has been devised.

Reproduction and development

A live bat in the hand readily yields a reasonable amount of information about its age and reproductive status. The epiphyses of the hand and wing bones of young bats have not yet fused to the diaphyses of the bones, and the characteristically knobby appearance of the joints of fully adult bats is lacking. Thus newly fledged bats are distinguishable from the older population for several months, and a sample may usually be separated into young of the year and older individuals. Lactating female bats have obviously enlarged nipples, and gentle pressure results in the expression of some milk. Females in advanced pregnancy are also conspicuous, although some bats with a full stomach may appear remarkably gravid. Sexually active males, especially of temperate species, may have the epididymides so swollen with semen that they extend obviously into the interfemoral membranes. Likewise sexually inactive males display no obvious epididymis, and the testes are minis-cule. The males of some phyllostomids, molossids, emballonurids, and pteropids have enlarged cutaneous glandular areas when breeding, and these places may be marked by pigmented, stained or elevated patches of hair. When copulation is occurring freshly captured females may reveal the presence of sperm on a slide made from a superficial swab of the vaginal opening. Thus a great deal about reproductive activity can be learned from examination of a temporarily restrained bat.

Ecomorphology

Because of the difficulty of studying ecological relationships of bats, some chiroptologists have resorted to the ecomorphological approach. Ecomorphology is the study of the relationships between morphology and ecological behavior. The idea is an old one, but rigorous attempts to infer ecological relationships from morphological ones date only from the 1950s and 1960s with the work of Brown and Wilson (1956) on character displacement, of Hutchinson (1959) on uniform size ratios between coexisting species, and of Van Valen (1965) on the niche variation hypothesis. Still, these earlier efforts dealt with single characters. The field of numerical taxonomy, with its emphasis upon the use of numerous unweighted characters in studying morphological relationships between organisms, introduced ecologists to a battery of multivariate techniques which allow comparisons based upon the simultaneous use of dozens of morphological traits (Sokal and Sneath, 1963; Sneath and Sokal, 1973).

The first essay into multivariate ecomorphology for bats was that of Fenton (1972) who compared bat communities by means of bivariate plots of ratios of external measurements, thus depicting four variables in a two-dimensional graph. Fenton was able to describe similarities and differences in flight behavior using this methodology and to compare faunas on this basis. Findley (1973) compared communities of bats of the genus *Myotis* from five regions using Euclidean distance in 48-space as a measure of morphological diversity. Karr and James (1975) conducted the first extensive multivariate ecomorphological comparisons of bird faunas of different continental regions using principal components and canonical correlation analyses. Subsequently ornithologists have employed these methods to study various bird communities (Ricklefs and Cox, 1977; Ricklefs and Travis, 1980; James and Boecklen, 1984; Leisler and Winkler, 1985), and additional studies have dealt with deermice (Smartt, 1978), stream fish (Gatz, 1979), and lizards (Ricklefs, Cochran and Pianka, 1981). Multivariate ecomorphological treatments of bat communities include those of Findley (1976), Mortensen (1977), Findley and Black (1983), Humphrey, Bonaccorso and Zinn (1983), Schum (1984), and Willig and Moulton (1989).

Two methodologies commonly used by ecomorphologists are ordination and distance analysis. Ordination involves construction of a resemblance matrix, either a correlation matrix or a variance–covariance matrix, comparing the morphological characteristics under study. From this matrix, linear equations are formulated which relate the original variables according to requirements which vary depending upon the kind of ordination analysis. These equations are known as eigenvectors or simply as factors. Each organism receives a set of scores, one for each factor, determined by entering the values of the original variables for that organism into each equation. Each score summarizes information from all the variables measured for the study animals in a certain way. Two or three scores often summarize most of the morphological data from a set of organisms which may then be arranged, or 'ordinated' in a two- or three-dimensional depiction which often accurately portrays their relationships with respect to the measured variables. The faith of the ecomorphologist is that this depiction is also that of ecological relationships. And, as we will see in Chapter 6, there is some good evidence that that faith may be justified.

A variety of ordination techniques is available. The commonly used principal components analysis proceeds with the requirement that the first factor (or component) accounts for the maximum possible amount of variation in the dataset. The second factor accounts for the maximum possible amount of the remaining variation and it must be orthogonal (that is, uncorrelated) to the first factor. The third factor accounts for the maximum possible amount of remaining variation, and must be orthogonal to factors 1 and 2, and so on. Other ordination methods are built upon other requirements. The relationship of each original morphological variable to each factor is determined by the 'loading' of the variable on the factor, that is, the degree to which variation in that original variable is expressed in the given factor. Thus each factor can be interpreted with respect to its ecomorphological significance. Each factor has an eigenvalue, a multivariate analog of variance, which shows how much of the variability in the original dataset is expressed in that factor. Typically two or three factors express most of the variability in a dataset of measurements, however if the morphological variables are not measurements, more factors are required to express most of the variation. The summed eigenvalues for a set of components provide a measure of the total amount of variability in the dataset, and that can be used to compare overall variation of the animals in several groups.

In distance analysis, the organisms of interest are arrayed in *n*-dimensional space, *n* being the number of morphological variables measured. Then the Euclidean distance between every pair of organisms in this space is determined, and the values entered as elements of a symmetric distance matrix. Since the distance increases as more morphological variables are added, either the same variables must be evaluated for each group of organisms, or each distance must be divided by the number of variables, giving the average distance.

In ecomorphological, as well as systematic, work the original data may be transformed in several ways, depending upon the requirements of the study. For example, if a set of measurements is taken from a bat, some measurements will be very large, such as head and body length, and some will be very small, such as foot or tooth length. The latter may be just as important ecologically as the former, but would contribute very little to the distance between different species, while the larger dimension would influence the distance disproportionately. To overcome this problem, the measurements may be standardized, that is, converted to Z-scores, by being expressed in terms of standard deviation units. Thus a large-footed bat would become just as distinctive as a large bat. An effect of standardization is to diminish the importance of absolute size. Of course, this may not correspond to ecological reality. The other common data transformation in ecomorphology is log-transformation. This has the advantage of making log-normally distributed data more normal, and of reducing the importance of very large measurements. Neither are required in ordination or distance analysis. However log-transformation does confer an advantage in principal components analysis: if it is used in conjunction with a variance–covariance matrix, the resulting combinations of character loadings on the components can be interpreted as ratios, and hence give information about shape of the animals (Mosiman and James, 1979). For example, if ear length loaded positively, and body length negatively on a given component, a bat which received a high score on that component would be relatively long-eared. In practice, results using standardized and log-transformed data are qualitatively similar.

Summary

The ecologist who wants to find out what bats are doing faces more logistical problems than does the student of birds, fish, lizards, primates or large grazing mammals. Even with a battery of electronic aids and a good deal of determination the rate at which ecological data accumulate is frustratingly slow. One may wonder, given the briefness of a professional life span, why any ecologist willingly selects bats as model organisms. Part of the explanation may be that the greater the mystery surrounding an organism the more appealing it becomes to some naturalists. Beyond that, however, the potential benefits suggested by the distinctive characteristics of the chiroptera (Chapter 1) have beguiled many investigators into believing that the costs of pursuing bat ecology may be worthwhile.

Some, frustrated by methodological problems, have turned to ecomorphology to supplement the understanding gained by more conventional methods.

4 · *A sampler of bat communities*

There are not many studies of bat communities. Bat ecologists tend to concentrate on autecological approaches or other endeavors with more clearly defined methods and goals. Most investigations that treat entire communities are descriptive and general in nature, not designed to test specific hypotheses or to answer specific questions. Nonetheless, these community-wide treatments provide a good deal of insight into the way bats are arranged in syntopic assemblages. Here I review some of this work. The goal is to accumulate information that may be useful in our later deliberations about patterns and processes in bat communities. I have used some of the information gleaned from these studies to construct morphograms of communities, depictions of the bats in ecomorphological space, that will help to conceptualize the ecological diversification and relationships of the component species. The methods and rationale used in this approach are dealt with in Chapter 3.

Many temperate-zone studies, especially those carried out in Europe, involve cave-dwelling, hibernating species. Most of the information provided by these efforts comes from bats captured in hibernacula, and may not reflect the composition of the local summer population of foraging animals. Tropical studies, especially in the neotropics, tend to emphasize bats captured in mist nets or harp (Tuttle) traps. Estimates of community structure based upon these data reflect several biases. Frugivores readily enter nets, while insectivores are difficult to net but rather easy to capture in harp traps. In a tropical forest, the sampled bat population is predominately that of the understory, while canopy foragers and species that fly above and away from the forest are more difficult to take. In arid regions, samples may be heavily biased in favor of species that visit water sources. Population studies that utilize sonic detection are limited to species with high intensity cries, with the result that the whole array of neotropical whispering bats and, of course, pteropids, are missed by this technique. Nonetheless, for regions where the bat fauna is composed chiefly of aerial insectivores, good quantitative estimates of the volant community may be obtained in this way.

In this review, the names of bats mostly conform to usage in Honacki, Kinman and Koeppl, (1982). I have in some cases changed the names used by the authors of the cited studies to reflect that reference.

Temperate North America

In the eastern part of North America one of the richest areas for bats is the Ozark Plateau in Missouri and Arkansas. Here a deciduous forest and a well-watered limestone region, with numerous caves, provides abundant hibernating, roosting, and foraging habitat. Within 100 km southwest of St Louis lie more than 200 caverns which support large wintering and summering populations. Up to 100 000 *Myotis sodalis* winter here, and as many as 50 000 *M. grisescens* spend the summer (LaVal *et al.*, 1977). La Val *et al.* (1977) studied the ways in which eight local species allocated foraging areas by marking the bats with chemoluminescent lights, then observing foraging from the ground or from a helicopter. *Myotis grisescens* tended to forage over water, while *M. sodalis* (mostly males were marked) worked in the forest canopy. *Myotis septentrionalis* foraged below the canopy and above the shrubs, while *M. lucifugus* (most of which migrate northward in the summer) fed in a variety of edge and forest situations. *Nycticeius humeralis* and *Pipistrellus subflavus* foraged over and near the water where they were captured, while *Lasiurus cinereus* and *L. borealis* flew high above the trees and pasturelands. The authors suggested that competitive displacement in habitat use was taking place, though no evidence of competition was presented. The data for this community suggest some allocation of foraging habitats. Lasiurines are seen as foragers in open airspace away from clutter and obstacles, *Nycticeius*, *Pipistrellus*, and *M. grisescens* favored the vicinity of water, while *M. septentrionalis* and *M. sodalis* were able to forage in the forest clutter.

Kunz (1973) studied the aerial bat communities at three sites in Boone County, central Iowa, where the predominant vegetation is deciduous forest. The sample consisted of 540 bats netted during 83 net-nights. Eight species were recorded, and time of capture of each individual was noted. Slight differences obtained between the three sites, and some well-marked differences between capture times of some of the species emerged. For example, the two species of *Lasiurus* differed in modal time of capture, *L. borealis* being caught most frequently in the first, and *L. cinereus* between 4 and 5 hours after sunset. *Myotis lucifugus* and *M. septentrionalis* differed in that the latter displayed a bimodal and the former a unimodal activity pattern. The two commonest species, *L. borealis* and *Eptesicus fuscus*, did not differ significantly in time of capture. Though differences in capture time may suggest temporal allocation of the foraging habitat, Kunz cautioned that proximity to the roosts of refuging species may influence the capture times at foraging sites.

Mumford and Cope's (1964) survey of the bats of Indiana provided some baseline information against which to compare community studies in the deciduous forest region of the eastern United States. Twelve species were listed, and maps provided showing the occurrence of each species in the state by county. In Table 4.1, the percentage of the 91 counties in which each species occured is shown. As in the Kunz study, the two commonest species are *Eptesicus fuscus* and *Lasiurus borealis*. *Eptesicus* roosts in tree hollows and under bark, as well as in buildings, while *Lasiurus* seeks shelter in foliage. Bats requiring caves as shelters, such as *Plecotus* and *Myotis grisescens*, are limited to a few counties. Generalist species which forage in clearings, forest edges, and over water for aerial prey and which can make use of the roosts commonly available in a forested region dominate the fauna. In all three eastern communities, as in the Russian and Scandinavian ones cited later in the Chapter, most bats are swift, direct fliers. Broad-winged species with slow, erratic or hovering flight are in the minority.

The richness of southwestern North American bat faunas has lured a number of chiroptologists, and the Southwest has been the site of some of the most intensive studies of local bat assemblages. The Mogollon Highlands of southwestern New Mexico and southeastern Arizona are especially well-favored, and two workers, Clyde Jones and Hal Black, have concentrated their efforts there. Jones (1965) examined 1595 bats, mostly from Socorro, Grant, Sierra, and Catron Counties, in west central New Mexico, and adjacent Greenlee County, Arizona. He recorded 20 species, and analyzed his data in three vegetative zones: (1) evergreen forest (= spruce–fir, mixed coniferous, and ponderosa pine forest), 18 species; (2) evergreen–deciduous woodland (= pinyon–juniper and oak woodland), 16 species; and (3) xeric shrub–grassland, 10 species. Jones recorded capture times, and showed some significant differences between species. His results were based upon parametric tests. If non-parametric comparisons as used by Kunz (1973) had been employed, the reported differences may not have proved to be significant. That 17 of these species may coexist at one site was shown by Black (1974) who sampled intensively at Nogal Canyon, at 2440 m elevation, in the San Mateo Mountains, Socorro County, New Mexico. The diverse topography of this station supports a mixture of pinyon, juniper, ponderosa and mixed coniferous forest. Black's chief goal was to study the food consumption of the bats and the composition of the flying nocturnal insect fauna. Insects were captured in blacklight traps, and food used by bats was estimated from examination of fecal pellets produced by the bats when they were held briefly after capture. Bats were netted at two major sites over water: a small concrete stock tank in a forest clearing, and a large earthen tank in open ponderosa–woodland transition. A major finding was that most bats of Nogal Canyon could be identified as beetle or moth specialists. In addition, Black categorized the bats with respect to use of four foraging areas: (1) between, within, and below the canopy, (2) open air, (3) terrestrial, and (4) over water. The first category included species that later students have referred to as gleaners as well as those identified as open-air foragers, and the terrestrial category comprised species frequently referred to as gleaners also. As in the earlier studies, the generalist *Eptesicus fuscus* was a common species, exceeded in abundance only by *Tadarida brasiliensis*, individuals of which gathered from a wide area to use the local water sources.

At 2600 m on the western slope of San Francisco Mountain, Coconino County, Arizona, Richard Warner (1985) studied the food habits of 10 of the species recorded by Black and Jones. Warner's study site lay in a meadow in ponderosa and aspen forest. Using Black's methods of fecal pellet analysis as well as the study of stomach contents, he confirmed the existence of a beetle-eating category, but noted that beetle-eating bats also fed heavily upon moths, while a few moth-eaters rarely ate beetles. Warner thus regarded moth-eaters as specialists and beetle-eaters as generalists. He was impressed with evidence of opportunism in feeding, rather than of competition and specialization.

Fenton and Bell (1979), while not conducting studies of entire southwestern bat communities, did provide data on the foraging techniques of some of the southwestern species of *Myotis*. *Myotis volans* was confirmed as a fast, direct-flying aerial forager, fixing on prey up to 10 m away. It was never seen gleaning. *Myotis auriculus* was seen to be mostly a gleaner, specializing on moths. *Myotis californicus* made many changes of direction in its foraging, with many feeding buzzes, and multiple attempts to capture prey.

Big Bend National Park, Texas, supports 18 species of bats which were the subject of a 5-year study by David Easterla (1973). He recorded bats in five plant communities as follows: (1) river floodplain–arroyo 550–1200 m, 16 species; (2) shrub desert, 550–1100 m, 15 species; (3)

sotol–grassland, 1000–1700 m, 3 species; (4) woodland, 1100–2400 m, 13 species; (5) moist Chisos woodland (cypress, pine, oak), 1800–2200 m, 13 species. All but the last of these habitats are very open from the standpoint of a flying bat, generally more so than the New Mexican and Arizonan stations mentioned. Easterla captured a total of 4807 bats during the study. Of these, 3549 were banded and released, and 106 were recovered, a return rate of 0.03%. The Big Bend fauna is dominated by cave–cavern–crevice roosting forms (*Antrozous*, *Leptonycteris*, *Tadarida brasiliensis*, *Nyctinomops macrotis*, *Plecotus*, *Myotis thysanodes*, *Pipistrellus*, *Mormoops*) and by species specialized for foraging over long distances in the open air (molossids generally).

The composition of the temperate North American communities discussed above is summarized in Table 4.1. The species involved are arrayed on two morphological principal components in Figs 4.1 and 4.2. All species were analyzed together with species from the Palearctic, Neotropical, Ethiopian and Oriental regions. Eight standardized morphological characters were used in a principal components analysis based on a character correlation matrix. The same scales are used in each morphogram. Western communities occupy more ecomorphological space than eastern ones. Space is extended by the addition of specialized hunters such as plecotines and molossids. Smaller communities, those of Iowa and Indiana, consist mostly of small bats with intermediate-sized ears and tails which suit them for aerial foraging in somewhat cluttered or in open airspace, and for some gleaning. An increase in species-richness accompanies increased availability of roosts as Humphrey (1975) has shown. Forested regions lacking cliffs, caverns, and caves, support fewer species, and those that do occur are known to use trees as daytime roosts in summer. Mountainous or broken topography with opportunities for roosting in crevices, cliff faces, caverns, and caves support richer communities.

The Palearctic region

In reviewing the fauna of the European part of the former Soviet Union, Strelkov (1969) grouped bats into migratory and non-migratory kinds. The latter comprise the permanent bat fauna of central and northern Russia, and include *Myotis dasycneme*, *M. daubentoni*, *M. mystacinus*, *M. nattereri*, *Plecotus auritus*, and *Eptesicus nilssoni*. These species hibernate in caves and cave-like shelters in central and northern Russia and in the Ural Mountains. Two are water-surface foragers, two are gleaners, one

Table 4.1 *Bat faunas of the United States discussed in the text*

Columns show presence (Missouri), number of individuals taken (Iowa, New Mexico, Texas) or percentage of counties in which species has been taken (Indiana).

	Missouri	Iowa	Indiana	Nogal Canyon, New Mexico	Big Bend, Texas
Gleaning insectivores					
Euderma maculatum				5	54
Idionycteris phyllotis				5	
Plecotus townsendii				1	496
Plecotus rafinesquei			1		
Antrozous pallidus				13	1113
Myotis septentrionalis[a]	X	64	7		
Myotis thysanodes				10	400
Myotis auriculus[a]				13	
Myotis evotis[a]				15	
Forest and clearing aerial insectivores					
Eptesicus fuscus[a]	X	243	20	195	68
Lasionycteris noctivagans[a]		52	5	23	
Pipistrellus subflavus[a]	X	2	12		
Pipistrellus hesperus				9	319
Nycticeius humeralis[a]	X	2	6		
Myotis sodalis[a]	X		6		
Myotis californicus[a]				X[b]	34
Myotis volans[a]				38	6
Myotis liebii				25[b]	1
Mormoops megalophylla					72
Water-surface foragers					
Myotis grisescens	X		1		
Myotis autroriparius			1		
Myotis yumanensis				20	384
Myotis lucifugus[a]	X	27	15	1	
Myotis velifer					61
Open-air aerial insectivores					
Lasiurus borealis[a]	X	124	20		
Lasiurus cinereus[a]	X	26	6	74	14
Tadarida brasiliensis				408	562
Nyctinomops femorosacca					89
Nyctinomops macrotis					411
Eumops perotis					91
Nectarivores					
Leptonycteris nivalis					632

Notes:
[a]Known to use trees as roosts.
[b]In the New Mexican study, *Myotis californicus* and *M. leibii* were not differentiated.

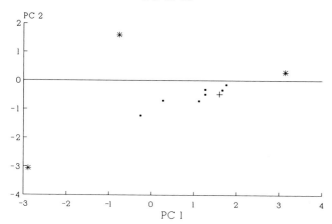

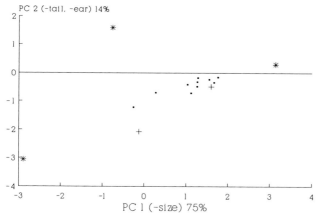

Fig. 4.1. Morphograms of bat communities from Iowa and Indiana discussed in the text. Each point represents one species. All species shown here and in Figs. 4.2–4.5 were analyzed together, and each morphogram ordinates them on the same scales. Larger communities occupy more ecomorphological space. Outliers are the extreme species from the world fauna. (· aerial pursuer; + hovering gleaner; * outlier)

Nogal Canyon, NM

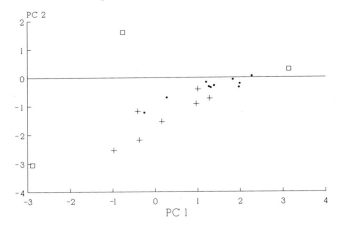

Big Bend, Texas

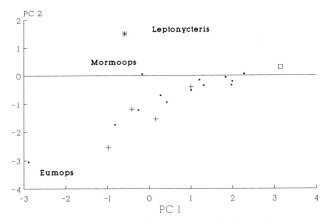

Fig. 4.2. Morphograms of bat communities from New Mexico and Texas discussed in the text. (· aerial pursuers; + hovering gleaners; * *Leptonycteris*; □ outliers)

specializes in forest clearings, and one forages at forest edges and in open areas.

To these in summer may be added the migrators *Nyctalus leisleri*, *N. noctula*, *N. lasiopterus*, *Pipistrellus pipistrellus*, *P. nathusii*, and *Vespertilio murinus*. Species on the second list migrate to the southwest to seek hibernacula. All seem to avoid caves and to seek instead sites in trees or buildings, places that would be too cold in the northern winter. Most possess aerodynamic configurations suiting them for sustained rapid flight, and are morphologically constrained to forage in relatively uncluttered areas. Thus the climate of much of the European part of the former USSR limits the composition of its summer bat communities to about 12 migrating or cave-hibernating species.

Many of the same species occur in southern Scandinavia. For that fauna, Baagoe (1987) predicted flight characteristics on the basis of flight morphology, and found that field observations and laboratory tests confirmed most predictions quite well. Species were arrayed on a spectrum from those with narrow wings, and fast, direct flight, which hunt well away from obstacles, such as *Nyctalus* and *Vespertilio*, to those with broad wings, and the capability for slow, maneuverable flight, which hunt around and among obstacles, such as *M. nattereri*, *M. bechsteini*, and *Plecotus*. *Myotis daubentoni* and *M. dasycneme* spent nearly all their time flying low over water. *Eptesicus* and *Pipistrellus* species were intermediate in speed, and often flew closer to obstacles than *Nyctalus* and *Vespertilio*, though not so close or so slowly as the large-eared kinds. The migratory species on the Russian list are those exhibiting fast and direct flight, and the foraging–trophic categories of Table 4.2 are in general supported by Baagoe's results.

A number of investigators have studied the hibernating bat communities of the limestone caves in South Limburg, Netherlands. Bezem, Sluiter and Van Heerdt (1960, 1964) gathered data on the summer and winter roosting sites of eight Limburg species. These bats differed in their preferences for outer or inner regions of the hibernacula, and these differences corresponded to the northern geographic limits of the several species in Europe. More northerly species, such as *Myotis mystacinus*, preferred outer cave regions. Southern species, such as *M. emarginatus*, preferred the deeper parts of caves where constant and somewhat warmer temperatures prevailed. Summer roosting sites also varied among the species, and to some extent reflected the winter preferences. Daan and Wichers (1968) studied winter roosting behavior throughout a hibernating season in one South Limburg cave. They noted that some species changed areas in the cave as the winter progressed, but confirmed the relationships shown by Bezem *et al.* (1964), and added the observation that northern species hibernating nearer the entrances to caves also showed more activity during hibernation and hibernated for a shorter period than southern species. Thus the bats of this Dutch community allocate summer and winter roosting sites, and as Baagoe

(1987) has shown, forage in different ways in different places. A majority of the Limburg species belong to the gleaning insectivore category.

It is clear that hibernating bats in Europe exhibit species-specific roosting preferences. In Poland, Bogdanowicz (1983) observed non-random associations and avoidances among hibernators, but concluded that these were driven proximally by different roost–site preferences and not by interspecific interactions.

The bats of the former German Democratic Republic (GDR) have been treated by Schober et al. (1971), who listed 17 species and provided dot maps of distribution as well as complete locality lists. There are many records in southern GDR where mountainous and forested terrain provide roosting and hibernating opportunities, but relatively few on the North German Plain, where these resources are less common.

Kowalski (1953) reported on 17 Polish species of which 11 were hibernators. As in the former GDR, the richest communities are in the mountainous southern part of the country. In the forest surrounding Pulawy, about 115 km southeast of Warsaw, Krzanowski (1956) recorded 16 species. All but three species (*Plecotus austriacus*, *Vespertilio discolor*, and *Eptesicus nilssoni*) were located in their summer roosts in bird boxes, tree hollows or wooden buildings. The forest and its resources can sustain most of the bats of northern Europe in summertime, so a factor limiting richness of summer communities must be the proximity of hibernacula and the ability of each species to make the flight needed to reach the forest area from the nearest hibernaculum. Forest and clearing aerial insectivores, with their faster, more direct flight, are the kinds that rank high in this ability.

The richer bat fauna of Czechoslovakia (Gaisler, Hanak and Klima, 1956; Gaisler and Hanak, 1969; Gaisler, 1975) consists of at least 27 species. Abundance and diversity of roosts, particularly of hibernacula, seem important in accounting for bat diversity here. The richness of the Czechoslovakian fauna can be appreciated by comparison with two of the richer states in the United States. Arizona and New Mexico have comparable diversity, 28 and 25 kinds of bats respectively, compared with Czechoslovakia's 27, but in an area 2.3 and 2.5 times larger, respectively, than that of the European country.

Figures 4.3 and 4.4 depict the European bat fauna, separated into several of the local communities discussed, arrayed on two principal morphological components. European communities are somewhat more diverse than North American ones, the space occupied by each tending to extend more symmetrically on both axes as a result of greater variation in size, ear, and tail dimensions.

Russia

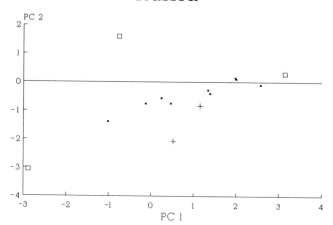

Limburg, Holland

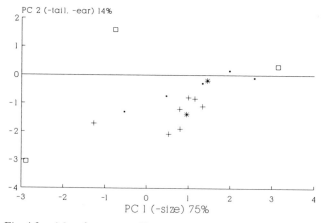

Fig. 4.3. Morphograms of bat communities from Russia and Limburg, Holland, discussed in the text. (· aerial pursuers; + hovering gleaners; ★ *Rhinolophus*; □ outliers)

In summary, the composition of European bat communities is strongly influenced by the availability of suitable hibernacula. Places lacking these resources, such as Scandinavia and much of the European part of the former USSR, are dominated by bats with fast, direct flight. Places with mountains or caves support a larger component of broadwinged, large-eared species, few of which seem to undertake extensive migratory journeys (Table 4.2).

Germany

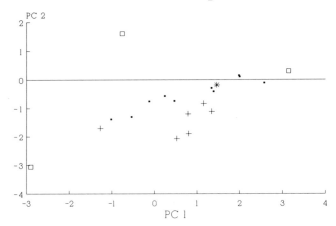

Czechoslovakia

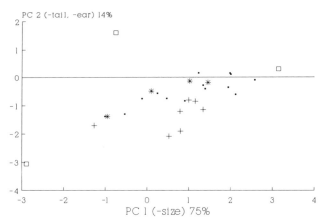

Fig. 4.4. Morphograms of bat communities from Germany and Czechoslovakia discussed in the text. (· aerial pursuers; + hovering gleaners; ⋆ *Rhinolophus*; ☐ outliers)

Tropical Africa

A classic among the earlier natural history studies of tropical African bats is that of Verschuren (1957) who spent two years investigating the bat fauna of Garamba National Park in northeastern Zaire (approximately 4°N). This 480 000 hectare reserve is mostly savanna and grassland, but also contains gallery forests and rocky outcrops which are important in providing roosting sites. Verschuren worked before bat nets were commonly used, with the result that a lot of his data came from bats taken at roosting sites or shot while foraging. Thirty-eight species were detected, and 1245 animals were examined.

Verschuren classified bats by roosting requirements. Externally roosting species, those that hang on tree branches or against trunks or rock surfaces, are seen to be rather unrestricted and to have the option of changing roost sites frequently. This category includes frugivores, *Lavia*, *Nycteris hispida*, *Hipposideros beatus*, and three species of *Taphozous*. Bats which roost inside shelters, cavities in trees, caverns, crevices, and other such places, are somewhat more restricted in distribution, and those which must hang in a pendant manner in their roosts require larger, and therefore less abundant, shelters. The latter category includes *Nycteris*, *Rhinolophus*, and *Hipposideros*. Internally roosting species, which roost with their bodies in contact with the substrate, enjoy a wider selection of sites because smaller cavities may be utilized. The vespertilionids and molossids fall into this category. In each roosting category, preferences for rocks, trees, or human habitations further subdivide species. Finally, division into forest- or savanna-inhabiting kinds delineates the allocation of shelters among Garamban bats.

The chief African forest study is that of Brosset (1966) who worked in the Haute Ivindo region of northeastern Gabon. The area is covered with undisturbed primary rain forest, and was investigated by Brosset for approximately 11 months. Over 1000 bats were examined, but three-quarters were of the species *Rousettus aegyptiacus*, *Epomops franqueti*, *Hipposideros commersoni*, and *H. caffer*, most of which were released after being identified. In all, 27 species were discovered. Most specimens were captured in mist nets at night, or were netted by hand at the diurnal roost sites. This study, together with that of Jones (1971) in Rio Muni, gives perhaps the best picture of a forest bat community from Africa. Six caves which were investigated contained four common species: *Hipposideros commersoni*, *H. caffer*, *R. aegyptiacus*, and *Miniopterus inflatus*. Culverts under roads sheltered *Nycteris grandis* and *N. hispida*, and those two species also used hollow trees together with *Hipposideros cyclops*. Fallen hollow logs were used by *Hipposideros beatus*. *Myotis bocagei* and *Pipistrellus nanus* occupied rolled banana leaves. Other foliage-roosting species included *Myonycteris*, *Megaloglossus*, *Epomops*, and *Hypsignathus*. A few kinds roosted in houses, and for some no roosts were ever found.

Jones (1971) listed 22 species for Rio Muni, mostly the same kinds as

Table 4.2 *Bat communities of European localities discussed in the text. Numbers in German Democratic Republic (GDR) column indicate percentage of bat records contributed by the species*

	European Soviet Union (former)	Scandinavia	Limburg	GDR (former)	Czechoslovakia	Pulawy
Gleaning insectivores						
Rhinolophus euryale			X		X	
Rhinolophus ferrumequinum			X		X	
Rhinolophus hipposideros			X	7	X	
Rhinolophus mehelyi					X	
Myotis emarginatus[a]			X		X	
Myotis nattereri[a]	X	X	X	5	X	X
Myotis bechsteini[a]		X	X	1	X	X
Myotis myotis			X	20	X	X
Myotis oxygnathus					X	
Plecotus auritus[a]	X	X	X	13	X	X
Plecotus austriacus[a]			X	8	X	X
Forest and Clearing Aerial Insectivores						
Myotis mystacinus[a]	X	X	X	6	X	X
Myotis brandti		X			X	
Vespertilio discolor[a]				1	X	X
Vespertilio murinus[a]	X[b]	X				
Eptesicus nilssoni[a]	X	X		1	X	X
Eptesicus serotinus[a]		X	X	8	X	X

Species						
Pipistrellus pipistrellus[a]	X[b]	X	8	X	X	X
Pipistrellus nathusii[a]	X[b]	X	2		X	X
Pipistrellus kuhlii[a]					X	X
Pipistrellus savii[a]						X
Barbastella barbastellus[a]	X	X	6	X	X	X
Water–surface foragers						
Myotis daubentoni[a]	X	X	7	X	X	X
Myotis capaccinii		X			X	X
Myotis dasycneme[a]	X	X	1	X	X	X
Open–air aerial insectivores						
Nyctalus leisleri[a]	X[b]		0		X	X
Nyctalus noctula[a]	X[b]	X	6	X	X	X
Nyctalus lasiopterus[a]	X[b]					
Miniopterus schreibersi	X			X		

Notes:

[a]Known to use trees as roosts.

[b]Present in the region in summer only.

those dealt with by Brosset, from an area of approximately the same geographic extent. Both authors treat seven megabats: the highly migratory *Eidolon*, the cavernicolous *Rousettus*, four foliage-roosting frugivores, and the nectarivorous *Megaloglossus*. Slow-flying pursuers, gleaners, or 'fly-catchers' included several *Nycteris*, *Rhinolophus*, and *Hipposideros*, as well as *Glauconycteris* (= *Chalinolobus*), and aerial pursuers included *Pipistrellus*, *Myotis*, *Miniopterus*, and *Tadarida*. Thus the forest community consisted of frugivores, and, among insectivores, emphasized slower-flying more maneuverable species (in contrast to the forest communities of the north temperate zones which are richest in open-air insectivores).

Findley and Black (1983) analyzed a syntopic assemblage of nine species of insectivorous bats from the vicinity of Lusaka, Zambia. The study was conducted over an 18-month period, and stomach content, and morphology of the bats were ordinated and compared. The community consisted of a cluster of morphologically and dietarily similar species and a smaller number of kinds which were more distinctive in both modalities. A three-dimensional model of the community in attribute space (Fig. 6.15) suggested that the more divergent kinds were also more variable intraspecifically in diet and morphology. The central cluster consisted of *Hipposideros caffer*, *Cloeotis*, and three species of *Rhinolophus*. These fed largely, up to 97%, on moths. The more divergent kinds were *Miniopterus schreibersi* and three species of *Nycteris*. These included more beetles in their diets, and the *Nycteris* also consumed non-volant prey such as lepidopterous and coleopterous larvae.

In Zimbabwe, Fenton *et al.* (1977) studied the foraging behavior and diets of 19 species of insectivorous bats sampled by mist-netting in the mopane and brachystegia woodland zone of Sengwa Wildlife Research Area. Of these species, most observations were of the vespertilionids *Pipistrellus nanus*, *Nycticeius schlieffeni*, *Eptesicus capensis*, and *Scotophilus viridus*. The pipistrelles foraged around buildings and in woods under the canopy. The other three species foraged among clumps of trees and above and below the canopy in more continuous woodland. All appeared to feed opportunistically on swarms of insects, although there were some differences in the proportions of moths and beetles taken, and perhaps in the species of beetles. These authors noted that flying insects seemed uncommon except in swarms, and had the impression that the bats mentioned did most of their feeding when swarms were encountered. Studies of insect remains in fecal pellets revealed that *Eptesicus*, *Nycticeius*, and *Scotophilus* relied chiefly upon beetles, and comparison of these results with insect samples from the woodland suggested that the bats selected beetles in excess of their proportional abundance.

An exemplary comparison of morphology, echolocation characteristics, flight abilities, foraging habitat, and diet, was carried out by Aldridge and Rautenbach (1987) on the insectivorous bats of Kruger National Park, South Africa. These authors established that bats foraging in the same habitats had similar diets, and that wing morphology and echolocation characteristics determined foraging habitat. Thus morphology provided a good guide to community trophic structure. They also showed that bat size was positively correlated with prey size. Small bats took only small prey, while larger bats took prey of a variety of sizes. The 26 syntopic species were divided into four morpho–sonic groups, but each species was distinctive, and the implication was that the members of the community allocated foraging habitats as a result of their morphological and sonic differences. Most bats belonged to group 1. These were small, less than 15 g, with low wing-loading, low aspect-ratio, low speeds, high maneuverability, and clutter-resistant calls. A few somewhat larger bats belonged to group 2. These had higher wing-loading and aspect-ratio, and were faster and less maneuverable. Group 4 bats, with high wing-loading and aspect-ratio, high speed, and reduced maneuverability included all the molossids and emballonurids. The large (90 g) *Hipposideros commersoni* was placed in a group by itself, with unusually low wing-loading and aspect-ratio for its size. Group 1 bats deal with small prey in cluttered habitats. Group 2 species must forage in somewhat more open areas in the woodland, while group 4 bats must utilize the open air above or away from the woodland. Bats assigned to these categories are identified by number in Table 4.3.

Among non-frugivorous African bats, three major trophic modalities are evident, and these represent intensifications of the modalities suggested by temperate-zone insectivores: (1) smaller, average-build, forest and clearing insectivores, (2) broader-winged, longer-eared gleaners, including sallying and fly-catching species, and (3) larger, faster, open-air pursuers.

Indo–Malayan and Australian regions

Community-level studies of bats in tropical Asia or in Australia are of recent vintage, and have been conducted using netting, trapping and sonic detection methods.

Near Chillagoe in northern Queensland, Fenton (1982a) used echolo-

Table 4.3 *Bat communities of Africa discussed in the text. Numbers following names refer to morpho-sonic groups of Aldridge and Rautenbach (1987)*

	Haute Ivindo	Rio Muni	Garamba	Kruger Park	Sengwa
Frugivores					
Eidolon helvum	X	X			
Epomophorus anurus			X		
Epomops franqueti	X	X	X		
Epomops gambianus					X
Epomops crypturus					X
Epomops wahlbergi					X
Hypsignathus monstrosus	X	X			
Megaloglossus woermanni	X	X			
Micropteropus pusillus		X	X		
Myonycteris wroughtoni			X		
Myonycteris torquata	X	X			
Rousettus aegyptiacus	X	X			X
Scotonycteris zenkeri	X				
Sallying and gleaning insectivores/carnivores					
Nycteris argae	X	X	X		
Nycteris nana	X		X		
Nycteris hispida	X	X	X		
Nycteris grandis	X		X		
Nycteris thebaica (1)			X	X	
Nycteris macrotis			X		X

Species	1	2	3	4
Nycteris woodi				X
Lavia frons		X		X
Rhinolophus denti		X		X
Rhinolophus fumigatus (1)		X		X
Rhinolophus alcyone	X		X	X
Rhinolophus clivosus				X
Rhinolophus landeri (1)	X		X	
Rhinolophus hildebrandti (2)			X	
Rhinolophus darlingi (1)			X	
Rhinolophus simulator				X
Rhinolophus swinnyi (1)			X	X
Hipposideros cyclops	X	X		X
Hipposideros abae		X		
Hipposideros caffer (1)	X	X	X	X
Hipposideros beatus	X	X		
Hipposideros commersoni (3)			X	X
Hipposideros ruber				X
Cloeotis percivali				X
Laephotis angolensis				X
Kerivoula lanosa	X			
Kerivoula argentata (1)			X	X
Myotis welwitschii				X

Forest and clearing aerial insectivores

Species	1	2	3	4
Eptesicus pusillus		X		
Eptesicus capensis (1)		X		X
Eptesicus rendalli		X		
Eptesicus hottentotus (2)			X	

Table 4.3 (*cont.*)

	Haute Ivindo	Rio Muni	Garamba	Kruger Park	Sengwa
Eptesicus somalicus (1)				X	X
Pipistrellus nanus (1)	X	X	X	X	X
Pipistrellus musciculus	X				
Pipistrellus rueppellii (1)				X	X
Pipistrellus rusticus (1)				X	X
Pipistrellus nanulus		X		X	X
Nycticeus schlieffeni (1)			X	X	X
Scotophilus nigrita			X	X	X
Scotophilus borbonicus (2)				X	
Scotophilus dingani (2)				X	
Scotophilus leucogaster					X
Myotis tricolor (2)				X	
Chalinolobus variegatus	X				X
Chalinolobus argentata		X			
Chalinolobus poensis		X			
Mimetillus moloneyi	X				
Water–surface foragers					
Myotis bocagei (1)	X	X		X	
Open-air aerial insectivores					
Miniopterus inflatus	X				

Species					
Miniopterus schreibersi		X			
Taphozous mauritianus (4)	X		X	X	X
Taphozous nudeiventris			X		
Taphozous perforatus	X		X		
Saccolaimus peli	X				
Chaerophon ansorgei			X		X
Chaerophon pumila (4)			X	X	
Chaerophon major			X		
Chaerophon aloysiisabaudiae	X				
Chaerophon nigeriae			X		X
Chaerophon chapini			X		X
Mops condylurus (4)			X	X	
Mops demonstrator			X		
Mops midas (4)			X	X	
Mops trevori			X		
Mops thersites		X			
Mops spurrellii		X			
Mops leonis	X	X			
Tadarida fulminans (4)				X	X
Otomops martiensseni			X		X

cation calls as a means of locating and identifying insectivorous bats. By counting the number of bat passes by different species in eight habitats, he was able to categorize bats as either those that sallied from perches to capture passing insects (*Nyctophilus bifax*, *Hipposideros diadema*, *Rhinolophus megaphyllus*) or those that hunted while in continuous flight. Of the latter group, Fenton recognized species which detected prey at distances of approximately 1 m (*Eptesicus pumilus*, *Nycticeius balstoni*), and those which detected prey at over 2 m (*Chalinolobus nigrogriseus*, *Taphozous georgianus*). Some differences in habitat use were documented as well. While Fenton was able to distinguish the calls of 12 kinds; of these, only nine could be assigned to their respective species.

The insect-feeding bat communities of the Kimberley mangrove stands in Western Australia were analyzed by McKenzie and Rolfe (1986). These woodlands occur as isolated patches which generally resemble one another in structure and floral composition. Twenty-two species of insectivorous bats inhabit the Kimberley subregion, but a maximum of eight were recorded in any given mangrove stand. Using flight morphology and direct observations, the authors were able to determine that, in each stand, the hunting bats were using different foraging zones, with relatively little overlap. Species with greater wing-loading and higher aspect-ratios (narrower wings), such as *Taphozous flaviventris*, *T. georgianus*, *Chaerephon jobensis*, *Mormopterus* cf. *beccarii*, *M. loriae*, and *Miniopterus schreibersii*, tended to forage above and outside of the mangal, while species with lighter loading and lower aspect-ratio (broader wings), such as *Eptesicus*, *Pipistrellus*, *Myotis*, *Nycticeius*, *Nyctophilus*, *Hipposideros*, and *Chalinolobus*, tended to forage within or close to the vegetation.

Crome and Richards (1988) compared the use by insectivorous bats of closed canopy forest and gaps in the forest caused by logging in an upland rain forest of the Mount Windsor Tableland in Queensland. The bats were detected and identified sonically, and activity in paired canopy and gap sites was compared. Habitat was predictable on the basis of the wing-loading and aspect-ratios of the bats. The authors divided the community into closed canopy specialists with low wing-loading and aspect-ratio (*Hipposideros ater*, *Rhinolophus megaphyllus*, *Nyctophilus bifax*, *Eptesicus sagittula*), gap specialists with high wing-loading and aspect-ratio (*Chalinolobus nigrogriseus*, *Scotorepens* (*Nycticeius*) *balstoni*, *Mormopterus loriae*, *M. beccarii*, and *Chaerophon jobensis*), and gap incorporators with intermediate morphologies (*Hipposideros diadema* and *Eptesicus pumilus*). Gap specialists were noted to use other open areas and not to be confined to gaps. Thus, as in the Kimberley study, the insectivore guild allocated foraging areas, and this allocation was predictable by morphology. Tight morphological packing among the gap specialists and among the canopy specialists suggested to the authors that vegetative structure, rather than competition constrained the structure of the community. Compositions of the Australian communities are tabulated in Table 4.4.

In a report on Malaysian rain forest microchiropterans Francis (1988) observed that forest bats, chiefly *Rhinolophus* and *Hipposideros*, had lower aspect-ratios than species that flew outside the forest, chiefly molossids and emballonurids. Similar differences were noted among vespertilionids, with forest-dwelling species having lower aspect-ratios than species which foraged in the open. Differences in wing-loading between forest and non-forest bats were not detected.

Francis (1990) also sought to compare the Malaysian rain forest bat communities with those of the neotropics, with special reference to the relative importance of frugivorous species. Intensive studies in peninsular Malaysia and in Sabah, northern Borneo, revealed that 15–18% of the species taken were frugivores, while 7–12% of the individuals fell into this category. By contrast, Francis noted that students of neotropical bats have reported that 34–48% of the species and up to 80% of the individuals taken in individual communities are frugivores. The Malaysian samples were made with mist nets and with modified Tuttle traps, and Francis commented upon the differential results obtained by the two methods. Fruit bats were commonly caught in nets, rarely in traps, while the reverse was true for insectivorous species. Most surveys were biased in favor of bats foraging at ground level. Nevertheless the preponderance of broad-winged species, those suited for slow, maneuverable flight, is marked. The Malaysian communities are listed in Table 4.5.

An exceptional contribution to our knowledge of megabat communities was recently made by Heideman and Heaney (1989). Working in submontane dipterocarp forest on Negros Island in the Philippines, these investigators handled over 200 mist-netted pteropids during the course of a study that extended over 6 years. Their work was concentrated in 302 ha of primary forest in the watershed of a small montane lake. Of the 13 species recorded, four (*Cynopteris brachyotis, Eonycteris spelaea, Macroglossus minimus,* and *Rousettus amplexicaudatus*) were common in agricultural habitats far from forest, and, within the forest, were mostly encountered in clearings. Three species (*Harpionycteris whiteheadi, Nyctimene rabori, Ptenochirus jagori*) were as common in clearings as in forest,

Table 4.4 *Australian bat communities discussed in the text*

	Chillagoe	Kimberley	Mount Windsor
Sallying and gleaning insectivores/carnivores			
Macroderma gigas		X	
Rhinolophus megalophus	X		X
Rhinolophus aurantius		X	
Hipposideros diadema	X		X
Hipposideros ater		X	X
Hipposideros stenotis		X	
Nyctophilus bifax	X	X	X
Nyctophilus arnhemensis		X	
Nyctophilus geoffroyi		X	
Nyctophilus walkeri		X	
Forest and clearing aerial insectivores			
Eptesicus pumilus	X	X	X
Eptesicus sagittula			X
Eptesicus douglasi		X	
Nycticeius balstoni	X	X	X
Nycticeius grayi		X	
Nycticeius cf. *sanborni*		X	
Pipistrellus tenuis		X	
Water-surface foragers			
Myotis adversus		X	
Open-air aerial insectivores			
Chalinolobus gouldi		X	
Chalinolobus nigrogriseus	X	X	X
Miniopterus schriebersii		X	
Taphozous georgianus	X	X	
Taphozous flaviventris		X	
Mormopterus loriae		X	X
Mormopterus beccarii		X	X
Chaerophon jobensis		X	X

Table 4.5 *Bat communities of East and West Malaysia discussed in the text.*
Numbers are of individuals captured

	Pasoh	Sepilok
Frugivores		
Balionycteris maculata	20	32
Chironax melanocephalus		3
Megaerops ecaudatus		1
Megaerops wetmorei	1	
Cynopterus brachyotis	5	43
Dyacopteris spadiceus		1
Penthetor lucasii	1	
Macroglossus minimus		1
Sallying and gleaning insectivores/carnivores		
Megaderma spasma	1	4
Nycteris javanica	1	5
Rhinolophus affinis	68	
Rhinolophus acuminatus		1
Rhinolophus borneensis		1
Rhinolophus refulgens	12	
Rhinolophus sedulus	3	14
Rhinolophus trifoliatus	31	19
Hipposideros armiger	1	
Hipposideros bicolor	73	
Hipposideros cervinus	33	198
Hipposideros dyacorum		42
Hipposideros diadema	11	11
Hipposideros galeritus	3	
Hipposideros larvatus	55	
Hipposideros ridleyi		2
Hipposideros sabanus	1	
Kerivoula hardwickii		1
Kerivoula intermedia		80
Kerivoula papillosa	21	78
Kerivoula pellucida	1	10
Kerivoula intermedia	18	
Phoniscus atrox	5	1
Murina aenea		1
Murina cyclotis	1	9
Murina rozendaali		1
Murina suilla		12

Table 4.5 (*cont.*)

	Pasoh	Sepilok
Forest and clearing aerial insectivores		
Harpiocephala harpia		2
Hesperoptenus blanfordi		18
Philetor brachypterus		3
Pipistrellus cuprosus		2
Myotis muricola	1	11
Myotis ridleyi	2	28
Myotis montivagus		10
Open-air aerial insectivores		
Emballonura monticola	2	
Emballonura alecto		3

but were absent at sites more than a kilometer from forest or forest patches. *Harpionycteris fischeri* was much more common in forest than in clearings, and was noted only in or very near forest or forest patches. The species of *Pteropus* and *Acerodon* were thought to fly above the forest and to forage in the canopy. They are under-represented in the sample which thus emphasizes bats which forage and fly in and below the canopy. Numbers and kinds of microbats taken were not reported. Numerical composition of the fruit bat community is depicted in Table 4.6.

The Neotropics

One of the earliest treatments of a syntopic bat community in the neotropics is that of Handley (1967) who took advantage of a netting program designed to study bird communities in the Belem area, eastern Brazil, to gather data on the differential use of ground level versus canopy by the resident bat community. Handley recorded 39 kinds of bats in 1157 individuals taken in nine nights of mist netting. A few kinds of bats were taken only in nets set in the canopy, up to 30 m above ground level, and a few were commonest there. Conversely some species were found only in ground samples. Thus early on, evidence came to hand that neotropical forest bats used different forest layers at different rates (Table 4.7).

McNab (1971) investigated the structure of some neotropical bat

Table 4.6 *Species, numbers, and principal habitat of fruit bats taken on Negros Island, Philippines*

Acerodon jubatus	1	Canopy and above
Cynopteris brachyotis	131	Open, clearings
Eonycteris robusta	6	
Eonycteris spelaea	67	Open, clearings
Haplonycteris fischeri	697	Mostly forest
Harpionycteris whiteheadi	140	Clearings and forest
Macroglossus minimu	335	Open, clearings
Nyctimene rabori	57	Clearings and forest
Ptenochirus jagori	603	Clearings and forest
Pteropus hypomelanus	1	Canopy and above
Pteropus pumilus	41	Canopy and above
Pteropus vampyrus	1	Canopy and above
Rousettus amplexicaudatus	43	Open, clearings

Source:
From Heidemann and Heaney, 1989.

communities, especially from Caribbean islands, as well as those from certain other parts of the world. He observed that these assemblages seemed to be structured on the basis of size and trophic category, and he constructed a niche matrix on the basis of these two variables. Rows were food categories, while columns were size. Each species was placed in a compartment of the matrix on the basis of its size and diet. In general, only one or two species occupied a cell of the matrix, and this was interpreted as being indicative of the allocation of food resources. A number of subsequent authors followed McNab's lead, and grouped species in their studies on the basis of similar matrices.

For example Fleming, Hooper and Wilson (1972) utilized a McNab matrix in their analysis of three Central American bat communities, two in Panama and one in Costa Rica. These authors noted multiple occupancy of matrix cells, especially by smaller insectivores and frugivores, but many of these kinds seem to be uncommon. Their study sites included: (1) Rodman Naval Ammunition Depot, 8 km W Balboa, Panama, a dry tropical forest habitat; (2) Fort Sherman Military Reservation 3 km W Cristobal, Panama, a moist tropical forest; (3) Finca La Pacifica, 4 km NW Canas, Guanacaste Province, Costa Rica, a riparian forest site. Thirty-four to 44 nights of mist netting were conducted over a period of a year (Table 4.8). Four basic reproductive patterns described

Table 4.7 *Vertical distribution of bats in the Belem area, Brazil*

	CO	CC	CG	GO
Rhynchonycteris naso				X
Cormura brevirostris	X			
Micronycteris sylvestris	X			
Micronycteris minuta				X
Micronycteris nicefori				X
Phyllostomus discolor				X
Phyllostomus elongatus				X
Phyllostomus hastatus			X	
Phylloderma stenops				X
Tonatia bidens			X	
Tonatia silvicola	X			
Trachops cirrhosus				X
Glossophaga soricina				X
Lonchophylla mordax				X
Lionycteris spurrelli				X
Choeroniscus minor			X	
Carollia perspicillata				X
Carollia subrufa				X
Rhinophylla pumilio			X	
Rhinophylla sp.		X		
Sturnira lilium		X		
Sturnira tildae			X	
Chiroderma trinitatum	X			
Chiroderma villosum	X			
Uroderma bilobatum			X	
Vampyrops helleri		X		
Vampyrodes caraccioli				X
Vampyressa bidens		X		
Ectophylla macconnelli				X
Artibeus sp. (small)		X		
Artibeus cinereus			X	
Artibeus jamaicensis			X	
Artibeus lituratus		X		
Artibeus sp. (medium)			X	
Desmodus rotundus				X
Desmodus youngi				X
Myotis albescens				X
Myotis nigricans				X
Myotis simus				X

Notes:
CO, canopy only; CC, commonest in canopy; CG, commonest at ground level; GO, ground level only.
Source:
From Handley, 1967.

included seasonal polyestry (most frugivores), seasonal monestry (a few insectivores), continuous activity with a short inactive period (*Myotis nigricans*), and uninterrupted reproductive activity (*Desmodus*).

Three Costa Rican stations used by the Organization for Tropical Studies were investigated in detail over a period of a year by LaVal and Fitch (1977). Finca La Selva lies at approximately 100 m in the Caribbean lowlands of Heredia Province and is covered with the Tropical Wet Forest of Holdridge's 1967 classification. Here bats were sampled during 261 net-nights and 76 trap-nights. Fifty-seven species and 1865 individuals were examined. Monteverde is located at 1400–1600 m in Premontane Moist Forest on the Pacific slope of the Cordillera Tileran in Guanacaste Province. Here 189 net-nights and 58 trap-nights yielded 905 individuals and 24 species. La Pacifica and Comelco Ranch are located at about 100 m in Tropical Dry Forest in the Pacific lowlands of Guanacaste Province. An unrecorded effort here resulted in the capture of 797 individuals distributed among 36 species. In all cases the sampling was chiefly by netting and trapping of bats flying near the ground. Construction of McNab matrices revealed multiple occupancy of many cells, especially by smaller frugivores and insectivores. Species and numbers taken at each station are included in Table 4.8.

Bonaccorso (1979) conducted a full-year study of the bat community of Barro Colorado Island, Panama Canal Zone. The 15 km² island lies in Gatun Lake on the Panama Canal, in the Tropical Moist Forest zone. Seventeen sampling stations were located in a 2 km² central strip of the island, 14 in closed canopy forest, and three along creeks. Each station consisted of four to six mist nets and one or two Tuttle traps. All night sampling was carried out 67 times in 1973, and partial night samples were conducted on 28 other nights. A total of 4376 net-hours and 1213 trap-hours were accumulated, and 2324 bats were captured in that year. A pilot study of 347 net-hours was conducted in 1971, when 282 bats were taken.

Bonaccorso categorized the bats into nine feeding guilds based upon food intake and foraging sites. Thirty-five species were detected in 1973, but three of them, *Noctilio labialis*, *N. leporinus*, and *Molossus molossus*, foraged in areas not sampled. A fourth species, *Vampyrum spectrum*, though taken in 1973, was extremely rare. Bonaccorso's guilds were described as follows, and bats in the neotropical communities tabulated in Tables 4.8 and 4.9 are categorized in a manner similar to this.

(1) Canopy frugivores. Species which forage for fruit in the canopy and subcanopy of the forest, more than 3 m from the ground.

Table 4.8 *Central American bat communities discussed in the text. Numbers are of individuals captured at each site*

Species	Station						
	Rodman	Sherman	La Pacifica	La Selva	Monteverde	Guanacaste	BCI
Forest and clearing aerial insectivores							
Rhynchonycteris naso			8	7		40	
Saccopteryx bilineata	3	11	7	17		12	X
Saccopteryx leptura				17		12	X
Peropteryx kappleri				1			X
Centronycteris maximiliani				4			X
Cormura brevirostris	7	4					
Pteronotus parnellii	3	26	4	27	7	269	X
Pteronotus davyi			5			10	
Pteronotus psilotus			1				
Pteronotus suapurensis					2	5	X
Pteronotus personatus						3	
Myotis nigricans	4	1		14	40		X
Myotis albescens			2	7			
Myotis elegans				20		4	
Myotis keaysi					226		
Myotis oxyotus					2		
Eptesicus andinus				1	1		
Eptesicus furinalis				1		2	
Rhogeessa tumida		2	1	1		9	X
Lasiurus ega					1		
Furipterus horrens				59			
Thyroptera tricolor		1			1		

Open-air insectivores

Balantiopteryx plicata						1	
Molossops greenhalli						1	
Molossus ater						6	
Molossus bondae	4			17			
Molossus coibensis	8	1					
Molossus molossus						7	
Molossus pretiosus						47	
Molossus sinaloae				76			X

Water bats

Myotis riparius			2	78			
Noctilio albiventris				1	4	44	X
Noctilio leporinus		4	8			25	X

Gleaning insectivores/carnivores/omnivores

Micronycteris sylvestris	1	2					
Micronycteris minuta	1	4					
Micronycteris nicefori				6			
Micronycteris brachyotis	1		3	3		5	X
Micronycteris hirsuta			2	4		1	X
Micronycteris megalotis			3	21	13	3	X
Micronycteris semidorum			1	1		1	
Barticonycteris daviesi				1			
Macrophyllum macrophyllum		2		3			
Lonchorhina aurita		3				3	
Tonatia bidens				15		1	X

Table 4.8 (*cont.*)

Species	Station						
	Rodman	Sherman	La Pacifica	La Selva	Monteverde	Guanacaste	BCI
Tonatia minuta				8			
Tonatia silvicola	3	15		1			X
Mimon crenulatum				10			X
Mimon cozumelae				16			
Phyllostomus discolor	8		46	1		7	X
Phyllostomus hastatus	15	2		1			X
Phylloderma stenops				1			X
Trachops cirrhosus	2	2	4	35		2	X
Chrotopterus auritus		1	1				
Vampyrum spectrum	1			4		4	X
Nectarivores							
Glossophaga commissarisi				81	34	10	
Glossophaga soricina	103	50	99	23	6	51	X
Lonchophylla robusta				12			
Anoura cultrata					1		
Anoura geoffroyi					13		
Choeroniscus godmani				3	8	5	
Hylonycteris underwoodi				15	7	3	
Lichonycteris obscura				1			
Ground-story frugivores							
Carollia castanea	30	48	3	302			X

Species							
Carollia brevicauda	325			461	53	75	X
Carollia 'perspicillata'		173	228	101			
Canopy frugivores							
Sturnira lilium			116		2		
Sturnira ludovici				1	287	22	
Sturnira mordax					8		
Uroderma bilobatum	73	292	1	17		1	X
Uroderma magnirostrum	1						
Vampyrops helleri	2	6		16			
Vampyrops vittatus					23		X
Vampyrodes caraccioli		9					
Vampyrodes major				1			
Vampyressa nymphaea	1	1		2			
Vampyressa pusilla	2	6		27			X
Chiroderma trinitatum	4	5					X
Chiroderma villosum		9	3	1		1	X
Ectophylla alba				38			
Artibeus watsoni		29		4			
Artibeus jamaicensis	356	483	328	111		24	X
Artibeus lituratus	56	22	31	16			X
Artibeus phaeotis	25	44	8	133		5	X
Artibeus toltecus					161		
Enchisthenes harti					1		
Centurio senex							X
Sanguinivores							
Desmodus rotundus	8	2	137	29			X

Note:
BCI, Barro Colorado Island.

Table 4.9 *South American bat communities discussed in the text. Numbers are of individuals captured at each site*

Species	Station		
	San Juan	Caatingas	Cerrado
Forest and clearing aerial insectivores			
Rhynchonycteris naso	19		
Saccopteryx bilineata	246		
Saccopteryx canescens	4		
Saccopteryx leptura	10		X
Saccopteryx sp.	1		
Cormura brevirostris	1		
Peropteryx macrotis	2	X	
Peropteryx trinitatus	1		
Diclidurus albus	7		
Diclidurus ingens	2		
Diclidurus isabellus	18		
Diclidurus scutatus	1		
Pteronotus parnellii	34		
Pteronotus davyi		X	X
Myotis albescens	28		
Myotid nigricans			X
Eptesicus brasiliensis	28		
Eptesicus furinalis	4		X
Lasiurus borealis	1		X
Lasiurus ega	8	X	X
Natalus stramineus			X
Furipterus horrens		X	
Open–air insectivores			
Molossops planirostris	228	X	
Molossops temmincki		X	X
Nyctinomops laticaudata		X	X
Neoplatymops mattogrossensis		X	
Eumops amazonicus	1		
Eumops glaucinus	68		
Eumops sp.		X	
Molossus ater	102	X	
Molossus aztecus	127		
Molossus molossus	1	X	X
Molossus sp.	37		
Promops nasutus	3		

Species	Station San Juan	Caatingas	Cerrado
Water bats			
Noctilio albiventris	212		
Noctilio leporinus		X	X
Gleaning insectivores/carnivores/omnivores			
Micronycteris hirsuta	1		
Micronycteris megalotis	23	X	X
Micronycteris microtis	1		
Micronycteris nicefori	1		
Micronycteris minuta	5	X	X
Micronycteris schmidtorum	9		
Glyphonycteris sylvestris	1		
Macrophyllum macrophyllum	2		
Tonatia bidens	1	X	
Tonatia brasiliensis	12	X	
Tonatia carrickeri	1		
Tonatia silvicola	3	X	
Mimon crenulatum	1	X	
Phyllostomus discolor	6	X	X
Phyllostomus elongatus	29		
Phyllostomus hastatus	255	X	X
Phylloderma stenops	6		
Trachops cirrhosus	29	X	
Chrotopterus auritus	3		
Vampyrum spectrum	3		
Nectarivores			
Glossophaga longirostris	16		
Glossophaga soricina	126	X	X
Glossophaga sp.	1		
Lonchophylla mordax		X	
Lionycteris spurrelli	18		
Anoura geoffroyi	2	X	X
Anoura sp. A	15		
Ground-story frugivores			
Carollia perspicillata	359		
Carollia sp.	855		

Table 4.9 (*cont.*)

Species	Station		
	San Juan	Caatingas	Cerrado
Canopy frugivores			
Sturnira lilium	626	X	X
Sturnira sp.	1		
Uroderma bilobatum	36		
Uroderma magnirostrum	208	X	X
Uroderma sp.	5		
Vampyrops helleri	154		
Vampyrops lineatus		X	X
Vampyrodes caraccioli	1		
Vampyressa bidens	22		
Vampyressa pusilla	1		
Chiroderma trinitatum	9		
Chiroderma villosum	494		
Chiroderma sp.	5		
Ectophylla maconnelli	1		
Artibeus concolor	10		X
Artibeus fuliginosus	48		
Artibeus harti	1		
Artibeus jamaicensis	306	X	X
Artibeus lituratus	61	X	X
Artibeus sp. A	6		
Ametrida centurio	1		
Sphaeronycteris toxophyllum	10		
Sanguinivores			
Desmodus rotundus	113	X	X
Desmodus youngi	5		
Diphylla ecaudata		X	

(2) Ground-story frugivores. Species which forage on fruits of shrubby ground-story plants, 0 to 3 m above ground level.

(3) Scavenging frugivores or 'juicers.' Feed mostly on very soft ripe fruit and/or over-ripe fruit.

(4) Nectar–pollen–fruit–insect omnivores. Forage for pollen and nectar when available during the dry season and for fruits and insects at other times (nectarivores in Tables 4.8 and 4.9).

(5) Sanguinivores. Feed on blood of mammals and birds.

(6) Gleaning carnivores. Forage for small animals (arthropods or verte-brates) that are perched or moving on vegetation or the ground.
(7) Slow-flying hawking insectivores. Forage for flying insects in small openings beneath or in the forest canopy or over streams (forest and clearing aerial insectivores in Tables 4.8 and 4.9).
(8) Fast-flying hawking insectivores. Forage for flying insects above the forest canopy or in very large open spaces such as pastures (open-air insectivores in Tables 4.8 and 4.9).
(9) Piscivores. Forage for fish or aquatic invertebrates just above the surface of lakes and large streams (water-surface foragers in Tables 4.8 and 4.9).

A matrix was provided arraying the members of each guild by weight. Although cells were not constructed for this treatment, clump-ing of species, especially within the slow-flying insectivore guild was evident. Bonaccorso commented extensively on differences he observed among members of each guild, including marked differences in abundance.

In the only community study carried out on mainland South Amer-ica, Willig (1983) compared the bats in two adjacent habitats in north-eastern Brazil, and categorized species as aerial insectivores, molossid aerial insectivores, foliage-gleaning insectivores, piscivores, nectari-vores, omnivores, frugivores, and sanguinivores. Over 5000 bats were captured, mostly in mist nets set at ground level, and 38 species were recorded for the two adjacent study areas. One area, the Caatingas biome, was richer in species, especially in the diversity of insectivores, presumably because of its greater topographic diversity and rocky outcrops which provided mesic enclaves (and probably roosts). The Cerrado biome, without this diversity, also was more depauperate in its bat community (Table 4.9).

In a later paper, Willig (1986) questioned the utility of guild-size niche matrices, concluding that they may obscure more than they reveal because numerous cells are occupied by several species while a majority are empty. He also noted that inhabitants of adjacent cells may resemble one another more than they do cell-mates.

Fleming (1986) reviewed the structure of several neotropical main-land and Caribbean island phyllostomid faunas searching for general rules governing the assemblage of neotropical bat communities. He noted that the structure of the species-rich communities closely resem-bled in taxonomic, trophic, and morphological characteristics the overall neotropical phyllostomid fauna. The individual communities generally contained clusters of morphologically similar species that tended to be less common and more specialized than the handful of common and broad-niched core species that are found in most communities. As species-richness diminished along rainfall and elevational gradients, specialist species tended to drop out of communities, while generalists persisted. Fleming concluded that mutualistic relationships with plants appear to be more important in structuring phyllostomid communities than do competitive interactions.

The work of Handley (1976) on Venezuelan mammals, though not involving studies oriented toward community structure, contains more data on syntopic neotropical communities than does any other available publication. As a result of the intensive collecting efforts reported by Handley, more kinds of bats are known from Venezuela than from any other area of similar size in the world. Among Venezuelan provinces and territories, the Federal Territory of Amazonas holds the record, 98 species, although Bolivar is close behind with 92. Bats were collected at about 10 sites in Amazonas. Mammals in the vicinity of the village of San Juan, 163 km ESE Puerto Ayacucho, at 155 m elevation, on the Rio Manapiare, were studied by Merlin D. Tuttle and Fred L. Harder from June 25 to August 3, 1967. There, with extensive help from native collectors, these workers obtained 5632 mammals, of which 5102 individuals were bats belonging to 78 species. Collections were made within 30 km of San Juan. The village lies in the Ventuari Basin, a plain nearly surrounded by high, forested mountains. There are many streams and lagoons, and the area is subject to extensive seasonal flooding. During the period of study the district near San Juan was about 95% flooded. The basin supports continuous evergreen forest (Tropical Humid Forest of Holdridge, 1967) with isolated savannas, containing scattered palms and bands of forest along streams. Many snags standing in lagoons, dead, often rotten and hollow, served as bat roosts. Most of the forest is undisturbed except for clearings made by the Indian population for gardens and villages.

In Fig. 4.5, 68 of the 78 species are depicted in a community morphogram ordinated on the same scales as are the temperate-zone communities. Larger species appear toward the left-hand side of the figure, with *Vampyrum spectrum* being outstandingly large compared to all other neotropical bats. Note that distribution on this axis is left-skewed, with a majority of species located toward the small end. This is a nearly universal result in principal component (PC) ordinations of

San Juan de Manapiare

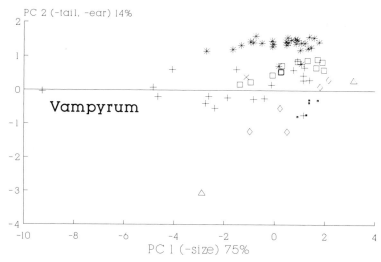

Fig. 4.5. Morphogram of the bat community from San Juan de Manapiare, Amazonas Territory, Venezuela. (■ vespers; + gleaners; ★ frugivores; □ emballs; × *Noctilio*; ◇ molossids; △ outliers)

mensural data, and such an arrangement is also manifested in the various MacNab matrices that have been constructed by other workers, all of whom note the multiple occupancy of cells by small insectivores and frugivores. The neotropical community differs from the temperate ones in the presence of a large number of frugivores (stenodermatines, glossophagines, carolliines) which, owing to short or absent tails and short ears, appear high on PC axis 2, and in general to shorter-tailed bats, such as emballonurids, which also appear above the zero-line on PC axis 2. Vespertilionids, the core of temperate-zone communities, play an insignificant role.

Myers and Wetzel (1983) analyzed the bat fauna of the Chaco Boreal, a savanna and grassland region lying in western Paraguay and adjacent Bolivia and Argentina, straddling the Tropic of Capricorn, and thus comparable latitudinally to Kruger National Park, South Africa. For purposes of comparison, those species examined by Myers and Wetzel from the Paraguayan Chaco (Departments of Alto Paraguay, Chaco, Nuevo Asuncion, Boqueron, and Presidente Hayes) are listed in Table 4.10. From this area of 247 000 km² lying between latitude 19° and 25°,

Table 4.10 *Bats from subtropical savannas and grasslands of the Chaco Boreal in Paraguay*

Water bats

Noctilio albiventris	51
Noctilio leporinus	59
Myotis riparius	10
Myotis simus	4

Gleaning insectivores/carnivores/omnivores

Tonatia bidens	2

Forest and clearing aerial insectivores

Peropteryx macrotis	1
Myotis albescens	193
Myotis nigricans	144
Eptesicus furinalis	49
Lasiurus borealis	2
Lasiurus cinereus	3
Lasiurus ega	79

Open-air insectivores

Molossops abrasus	6
Molossops planirostris	10
Molossops temmincki	80
Eumops bonariensis	229
Eumops dabbenei	9
Eumops glaucinus	5
Eumops perotis	4
Nyctinomops laticaudata	7
Nyctinomops macrotis	1
Promops centralis	1
Promops nasutus	5
Molossus ater	13
Molossus molossus	92

Frugivores

Sturnira lilium	1
Artibeus lituratus	2
Artibeus jamaicensis	1

Sanguinivores

Desmodus rotundus	20

Source:
From Myers & Wetzel, 1983.

these workers examined 1083 bats belonging to 29 species. Most of the individuals and species are aerial insectivores, mostly molossids adapted for foraging in open air. Comparison with the African grassland and savanna communities is intriguing: although about the same number of species and of trophic categories are represented, the sallying/gleaning category is essentially absent from the Chaco. The great neotropical preponderance in species and trophic categories seen in the comparison of neotropical with the African forest communities does not exist. Rather, south temperate grasslands and savannas in both continents are approximately the same in species richness.

Summary

Depauperate bat communities, located in the northern temperate regions, comprise chiefly aerial insectivores. These species hunt their prey on the wing, and, depending upon their size and upon the length of their ears, prefer more open or more cluttered airspace as foraging areas. In regions where suitable hibernacula are uncommon, faster species are more common than those better adapted to slower, more maneuverable flight, perhaps because their economical flight allows them to summer substantial distances from their hibernacula. The commonest north-temperate-zone bats are those that can utilize trees as roosts. As communities become more species-rich in more three-dimensional habitats, hovering and gleaning bats become more common, as do kinds specialized for high-speed long-distance flight, such as molossids. Finally frugivores appear. This progression involves an expansion of ecomorphological space, but in most communities a core of small insectivores, and perhaps frugivores, remains. Within this core there are many species that resemble one another closely in ecology and morphology. Bats do not seem to be hyperdispersed with respect to size and trophic category as implied by the McNab matrix.

5 · *Resource limitation and competition in bat communities*

Much of the literature of evolutionary ecology carries the implicit or explicit assumption that resource limitation and competition exist in natural communities, and a great deal of field data are interpreted in the light of this assumption. Concrete demonstration of either process is relatively uncommon. Although the difficulty of providing irrefutable evidence of competition among animals is notorious, a reasonable number of examples is available in the literature (Connell, 1983; Schoener, 1983). The distinctive biological properties of bats lead one to suspect that, in this group of vertebrates, if in any, these phenomena ought to be demonstrable. To what extent does the evidence support the existence of resource limitation and competition for bats?

The difficulties encountered in working with bats have thus far largely ruled out the use of experimental manipulation in attempted demonstrations of resource limitation and competition. Thus most of the evidence mustered below comes from comparative studies or from correlations between behavior and morphology. Of course, such evidence is not as satisfying as that emanating from carefully controlled experiments. Yet, many important advances in natural history have depended on just this kind of support.

Food

Both frugivorous and insectivorous bats adjust their reproductive cycles so that young are born during periods of food abundance. This has been abundantly documented for the diverse communities of Central America, for example by Fleming *et al.* (1972), and Bonaccorso (1979). On Barro Colorado Island, Panama Canal Zone, Bonaccorso (1979) noted that fruit is limiting seasonally, to the extent that certain frugivorous species left the area, were captured with empty stomachs, or switched to other diets during the late rainy season when fruit was scarce. In New Mexico, Black (1974) recorded a close correlation between the weight of

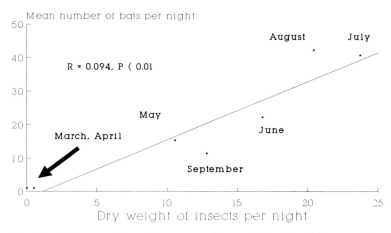

Fig. 5.1.   Mean number of bats captured per night in Nogal Canyon, New
Mexico, as a function of mean dry weight of insects captured per night. (After
Black, 1974.)

insects captured in light traps and the mean number of bats netted per
night at the same site (Fig. 5.1). In Rio Muni, West Africa, Jones (1972)
found that the abundance of pteropids was correlated with the abundance of fruit, and that some species of bats seemed to leave the area
when fruit was scarce. McNab (1982) believes that the chief factor
preventing most tropical bats from living in the temperate zones is lack
of food. Bat reproduction, occurrence, and abundance are related to
food abundance, and in that sense food is clearly limiting to the animals.

Syntopic species of bats by definition are different from one another,
not only in morphology, but presumably in every other way, including
in diet. Is there evidence that these differences in diet have been driven by
the requirements of coexistence? Here the evidence is conflicting.
Several workers have noted that insectivorous bats foraging in the same
places consume the same kinds of insects (Fenton and Morris, 1976;
Aldridge and Rautenbach, 1987). Fenton and Morris (1976) thought that
the insectivorous bats, mostly *Myotis* spp., in their arid study area in
Arizona were not specialists with respect to types of insects, but rather
were adapted to take advantage of insect concentrations whenever they
occurred, without reference to insect identity. In reviewing evidence for
resource partitioning, Fenton (1985) concluded that dietary specialization was rare in bats, and that most were generalists. Others have noted
some evidence of partitioning and specialization. Black (1974) grouped
insectivorous bats in his montane study site in New Mexico into moth

and beetle specialists. Findley and Black (1983) noted some specialization, chiefly on the basis of the volant or non-volant nature of the prey, in a community of nine insectivorous species in Zambia. In Kruger National Park, South Africa, Aldridge and Rautenbach (1987) found that, while there were significant differences in the diets of insectivorous species foraging in different habitats, species in the same habitat tended to have similar diets. These authors also noted a correlation between mean size of prey and predator size. They further observed that large bats captured prey of all sizes, while small species consumed only small insects. Bonaccorso (1979) reported a correlation between size of bat and size of fruit in the canopy frugivore guild of bats on Barro Colorado Island. Food niche overlap in this guild was a function of similarity in size of the bats, but apparent niche was often a function of sample size, so that obtaining larger samples from a species increased its niche breadth. McNab (1971) surveyed some neotropical and paleotropical bat communities and observed that, based on the supposition that food particle size and bat size are correlated, bat communities seemed to be so arranged that food was allocated on the basis of major category (fruit, insects) and within category on the basis of food particle size. Most subsequent studies do not support that observation. Withal, evidence of food resource allocation based upon studies of diet is equivocal. Fenton (1982b) concurred in this evaluation with reference to resources in general, but later (1985) admitted that this 'stance has been somewhat ameliorated by recent studies.'

Two studies seem to provide strong evidence of competition for food between syntopic bats. Husar (1976), in studying the gleaning insectivores *Myotis auriculus* and *M. evotis* in New Mexico, noted that at a station where the two were syntopic one favored beetles while the other specialized in moths. Where the two occurred allopatrically these differences disappeared (Fig. 5.2). Still more suggestive, in the allopatric situation, males and females diverged in diet in the same way, while in sympatry the sexual differences were not noted. In the Ivory Coast, Thomas (1985) studied the pteropids *Micropteropus pusillus* and *Epomops franqueti*. Both fed upon the fig *Ficus capensis*. During the dry season, when fruit availability is low, *Epomops* took all the fruit, and populations of *Micropteropus* dropped by 50%. After a discussion of the possibility of food competition between frugivorous vertebrates, Fleming (1979) accepted the idea that, in the absence of contrary evidence, competition for food resources has been the principal selective factor in determining the dispersion of ecological roles of frugivorous bats. But in a later reflection (Fleming, 1986) he abandoned the notion, and concluded that

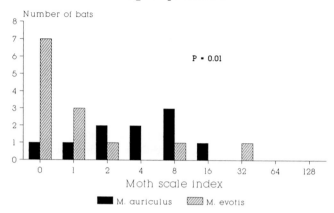

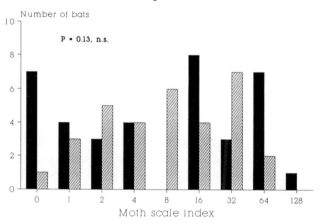

Fig. 5.2. Moth consumption, as indicated by counts of scales in fecal pellets, by *Myotis auriculus* and *M. evotis* in areas of allopatry and of sympatry. Where the two species coexist diet diverges. (After Husar, 1976.)

mutualistic relationships with plants were more important influences on frugivore community structure.

Foraging areas

While foraging areas themselves may not be limiting, foraging habitat specialization may result from competitive interactions between bats. A reasonable amount of direct and indirect evidence does indeed show that syntopic assemblages of bats may allocate foraging habitats.

Aldridge and Rautenbach (1987) in their studies of the Kruger National Park bat fauna, divided foraging areas into seven categories, then by tracking light-tagged bats, quantified the amount of use each species made of the different habitats. They observed that the insectivorous species fell into four categories with respect to habitat use: open foragers, woodland-edge foragers, intermediate clutter foragers, and clutter foragers. Each category contained from one to three common species. Occupancy of a category was determined chiefly by wing morphology, which in turn dictated the ability of the species to maneuvre in a cluttered airspace. Even though the bats using the same foraging strategy tended to eat the same foods, the use of different habitats tended to reduce food overlap.

In southern Finland, Lehmann (1985) observed that three common vespertilionids, *Myotis brandti*, *M. daubentonii*, and *Eptesicus nilssoni*, foraged in different habitats: in 'not too dense forest,' over water, and in open places, respectively.

Studying insectivorous bats in Queensland rainforests, Crome and Richards (1988) found that their species tended to be gap specialists or closed canopy specialists, with some species foraging in the closed areas but incorporating gaps in their foraging areas. Gap specialists tended to have high wing-loading and high aspect-ratios, while the opposite traits characterized closed canopy foragers. Within each category the species were closely packed morphologically.

In the Kimberley Mangroves of northern Western Australia, McKenzie and Rolfe (1986) found that predictions of the foraging niches of insectivorous bats based upon flight morphology were supported by field observations. At each site studied, the bat species allocated foraging niches with some overlap. Little overlap between species was observed in the morphogram of the community.

In the Krau Game Reserve of West Malaysia, 12 syntopic species of *Rhinolophus* and *Hipposideros* are found. Heller and Helversen (1989) recorded and analyzed the echolocation calls of these species. Dominant frequencies (those with the strongest amplitude) of the 12 species had a significantly more even distribution than predicted by chance. These investigators suggested that partitioning of prey or of communication channels was taking place.

Thus a small though impressive array of careful studies supports the thesis that syntopic bats allocate foraging habitats, that this allocation depends upon flight morphology and echolocation characteristics, and that habitat use can be predicted from morphology. Bats are thus physically and behaviorally adapted for different kinds of foraging areas. But does this mean that space itself is, or was, an important limiting resource for the animals or that space is used as a means of subdividing some other resource, such as food?

Roosts

The roosting ecology of bats has been well reviewed by Kunz (1982). Many bats have roosting requirements that are determined by physiological demands of the adults or young, by predation pressures, by sociological considerations, or by morphology.

In some cases, roosting requirements are so specific that the absence of suitable roosts precludes the occurrence of the species. For example, the neotropical disk-winged bat *Thyroptera tricolor* roosts in the rolled new leaves of *Heliconia* plants and is physically specialized for clinging to the smooth surfaces of leaves by the possession of adhesive disks on wings and feet. No alternative roost sites have so far been discovered for this species, and it is likely that occurrence of the plant limits the distribution of the bat. A number of species of phyllostomids of the subfamily Stenodermatinae construct roosts by modifying the structure of leaves of bananas, palms, or other plants, and may be similarly limited by the distribution of the plants.

Humphrey (1975) has shown that species richness and diversity of colonial nearctic bats are strongly correlated with an index of physical structure of the environment which includes contributions from topographic complexity, presence of trees, and human constructions. High bat diversity characterizes areas where all kinds of roost structures occur, whereas places with low bat diversity are also lacking in one or more roost types. In all likelihood, this generalization applies to all temperate regions. In the European region of the former Soviet Union, bat diversity increases toward the southwest where mountainous terrain provides additional roost possibilities (Strelkov, 1969), and in Europe generally the richest faunas are found where diverse topography, forests, and human constructions provide opportunities for shelter (Gaisler *et al.* 1956; Krzanowski, 1956; Egsbaek and Jensen, 1963; Schober *et al.* 1971; Gaisler, 1975).

The extent to which roosts are as limiting in tropical regions is less clear. Graham (1988), in a study of roosting ecology along an altitudinal gradient in Peru, came to the conclusion that there was no clear evidence that roosts there are limiting resources, or that there is competition for roosts. Many tropical forest bats roost in caves, but many others utilize tree hollows or foliage. Foliage roosts seem unlikely to be limiting unless a specific kind of plant, or stage in the development of a plant, is required, as is the case with *Thyroptera*. Cavities in trees could be limiting, and indeed Morrison (1979) has found that male *Artibeus jamaicensis* actively defend hollows that are required by the females for the bearing of young. By this defense, the males gain reproductive access to the females. The females themselves are the resource for the males, but their availability is limited by the appropriate roost site. Verschuren (1957) has described and analyzed the roosting sites chosen by bats in northeastern Zaire. These are very diverse and in some cases seem quite specialized. Nevertheless there is no clear suggestion that any of them may limit the occurrence of the bats.

In cases where different species of bats occupy the same major roost, for example a large cavern, it is often noted that each species seeks a different microhabitat within the larger structure. Bogdanowicz (1983), for example, noted non-random roosting associations among the species inhabiting various roosts in Poland. He believed, however, that this differentiation arose from differing roosting preferences rather than from interspecific interactions.

As noted in Chapter 4, summer bat communities in forested parts of northern Europe are constrained by the proximity of suitable hibernacula. This constraint affects not only the number of species but their morphology and trophic behavior as well. Similarly in temperate North America, alpha diversity and morphological and trophic structure are influenced by the proximity of roosts.

For some species there is clear evidence that the provision of artificial roost sites allows an expansion of their range. Fenton (1970) has shown this for *Myotis lucifugus*, which quickly takes advantage of man-made structures such as buildings and mines, and has seemingly increased its range in North America since European settlement.

In summary, there is clear evidence that roosts may limit the geographic occurrence of bats, but very little suggesting that competition for roosts may structure local bat communities.

Heat

Bats are chiefly tropical, and it is easy to imagine that they are progressively excluded from more northern regions by lower temperatures.

However, this is seemingly not the case. McNab (1982), in reviewing the physiological ecology of bats, has shown that, except perhaps for vampires, food supply, not temperature, limits the northward dispersal of tropical bats. In local situations, internal temperature may limit bat distribution. Many female temperate zone vespertilionids become homeothermic during the season of parturition and lactation. In Washington and Oregon, Thomas and Bell (1986) have shown that in some areas cloudy and rainy weather during that season inhibit insect activity and foraging by the bats. Male *Myotis*, which remain heterothermic in summer, are able to persist in such places because of their ability to become hypothermic when regular feeding is impossible. Females, however, cannot forage enough in such places to maintain homeothermy, and hence are limited to places that are less cloudy and rainy.

Water

On the basis of numbers of species and individuals, Carpenter (1969) thought that bats might be the most successful desert mammals. Happold and Happold (1988), however, pointed out that, although there are about equal numbers of bats and rodents in both North American and Australian deserts, chiropterans are outnumbered by rodents two to one in the driest parts of the Sahara, and three to one in the Namib. Thus, although these relationships may be affected by roost availability, the relative success of bats in deserts is open to question. Some bats may persist without drinking water for long periods in captivity (Happold and Happold, 1988). Whether they do this in the wild is not certain. By spending the day in appropriate roosts, and flying to water sources at night, bats may occupy some of the most formidable deserts.

There is a clear relationship between kidney structure and function and the degree to which bats inhabit arid environments. Geluso (1978) demonstrated a correlation between urine concentrating ability and the ratio of inner medullary zone to renal cortex. Bats, such as *Pipistrellus hesperus* and *Antrozous pallidus*, with the best-developed inner medullary zone, were able to achieve greater concentrations, and also were the species most common in the lower and more arid parts of the American Southwest. Studier *et al.* (1983) surveyed neotropical bats, and found that frugivorous species had the thinnest medullas, whereas species from arid zones were as predicted by Geluso (1978). Studier *et al.* (1983) found the highest correlations between diet and kidney structure, with frugivores having low, and animalivores high medulla to cortex ratios. In surveying the bats of Malawi, East Africa, Happold and Happold (1988) found wide variation in the kidney structure and function of bats inhabiting each ecological zone, and concluded that efficient kidneys are not essential for desert occupancy by the species they studied. That these authors considered 35% relative humidity to be low, however, may cause one to suspect that extreme desert conditions do not obtain in Malawi. Pteropids included in the Malawi study had a thin and undifferentiated medullary zone, and since these fruit bats are not known to drink in nature, it is assumed that their water needs are supplied through the diet.

In situations where surface water occurs in limited amounts and in restricted situations, huge numbers of bats may appear simultaneously over water holes. This has been noted especially in the arid western parts of the United States. Under these circumstances it is conceivable that physical access to the water surface may be somewhat limited, and indeed there is a little evidence that temporal partitioning of appearance at the water source may sometimes occur, but, in a comparison of watering times of 19 species in New Mexico, Jones (1965) revealed few significant differences.

In most parts of the world it does not appear that water limits bats in such a way that it has a significant impact on community structure.

Summary

There seems to be some evidence that competition has actually affected bat communities, but a also fair amount of conjecture that it has been involved. Much evidence illustrates that bats have differentiated in ways that result in resource allocation. In certain cases, resources demonstrably place a limit on chiropteran existence. But while competition may influence bat community structure, the available data do not provide strong support for the controlling nature of this factor.

The availability of suitable roosts does limit the richness and morphological and trophic diversity of bat communities in the temperate zones. Tropical communities may be similarly limited, but good evidence to support that possibility is not available.

6 · *Patterns in bat communities*

In this chapter I seek patterns in species richness, taxonomic, trophic, and morphological diversity, in packing, and in biomass and numerical abundance. Such patterns, if they exist, may suggest questions leading to a better understanding of the forces that have structured bat communities.

Patterns in species richness: where the bats are

Most bats live in the tropics. At latitudes north of 50°, there are very few kinds of bats, and north of 60° there are almost none. In the southern hemisphere, bat ranges extend to the limits of the major land masses, but bats fail to reach Antarctica. Within the temperate and tropical zones, bats display a classical increase in species richness as the equator is approached. This was clearly shown for North America by J. W. Wilson (1974) and for the entire New World by Willig and Selcer (1989) (Fig. 6.1). Figure 6.2, based upon fewer data points than Fig. 6.1, shows the expected trend for the Old World. However, when isograms for species richness, based on quadrats comparable to those used by Willig and Selcer, are plotted for the Old World (Fig. 6.3), it is seen that for the African continent the classical pattern is greatly complicated. First of all, the Sahara Desert results in a dip in species number. Then richness increases again, reaching a peak in equatorial East Africa in the savanna and grassland region. But the peak, between 60 and 70 species per 500 km² quadrat, is only about half of the New World maximum. The unique feature of the African cline is that richness does not peak in the equatorial rain forest. Indeed, quadrats located within that biome have fewer kinds of bat than do those in the surrounding savannas. This pattern will assume importance in our later search for causes. Many groups of quadrupedal terrestrial mammals are not more speciose in the tropics, and the trend for the class Mammalia is driven by the Chiroptera. The contrast between temperate and tropical regions is dramatic.

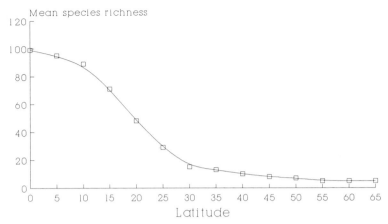

Fig. 6.1. Species richness of New World bats as a function of latitude. $R^2 = 0.87$ for the simple polynomial regression curve shown. Each point represents the mean richness for all of the 500×500 km quadrats at the given latitude. (After Willig and Selcer, 1989.)

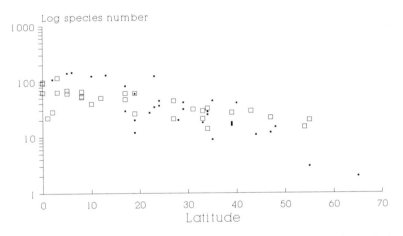

Fig. 6.2. The world relationship between latitude and species richness. Points represent political units and islands for which reliable data are available. Area varies widely among these geographic and political units, yet the strong correlation is still evident. Note that, in the tropics, New World stations are generally richer than Old World ones at comparable latitudes. (From various sources. ■ New World; □ Old World)

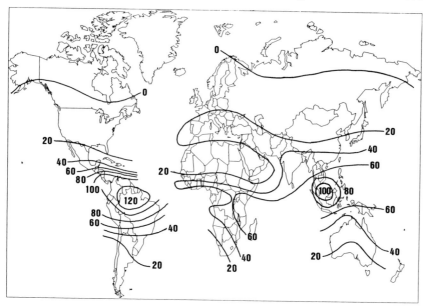

Fig. 6.3. Species richness of bats throughout the world. Isograms are based on
bat species richness in quadrats 500 km square, following the methodology of
Willig and Selcer (1989). Data points for the Old World are widely scattered, and
the isograms there are approximate. Note that in the New World and in Asia and
Australia richness increases more or less steadily as the equator is approached. But
a transect from northern Europe south through Africa shows no such regular
pattern. Instead, richness drops in the arid regions of north Africa and the Arabian
Peninsula, increases in the grasslands and savannas south of the Sahara, and then
drops again in the Central and West African equatorial rain forest regions. The
peak of African bat diversity is in the grasslands and savannas of equatorial East
Africa. The projection of the map is a special one, designed by my computer,
with the goal of fitting a maximum amount of the earth's land surfaces on an
8.5 × 11.5 in. page.

Handley (1976) records 146 species for Venezuela, almost twice the number recorded for the entire Palearctic region. Koopman (1970) has reviewed the zoogeography of bats on a world-wide basis.

Within the tropics there are three main centers of species richness, the American and the southeast Asian tropical rain forest regions, and the East African equatorial savanna region. By far the greatest richness is in the neotropics, centered over Amazonia, with perhaps the maximal species density occurring in the headwaters of the Amazon, in a crescent around the northern and western side of the basin. Only slightly behind are the rain forests of southeast Asia, centered over the Malay Peninsula and Borneo. For Africa, savanna bats, not forest species, dominate, reaching peak densities in Uganda, Kenya, and Tanzania.

To the community ecologist, the number of coexisting, syntopic species is the matter of greatest interest. Here the world record seems again to be in the neotropics, and to be held by sites in Costa Rica and Venezuela. At Finca La Selva, Heredia Province, in the Caribbean lowland rain forest of Costa Rica, 65 species have been recorded, and another 20 are assumed to occur based upon occurrence in similar or contiguous areas. At Rincon and at Corcovado Park on the Osa Peninsula, Puntarenas Province, in the Pacific lowlands of Costa Rica, 50 species have been taken, and it is suspected that another 30 occur (Wilson, 1983). In Amazonas Province, southern Venezuela, 78 species have been taken at San Juan, Rio Manapiare, 163 km ESE Puerto Ayacucho, and from 40 to 53 species have been taken at five other sites in that province (Handley, 1976). San Juan seems to hold the world record for bat alpha diversity. These numbers are not approached at paleotropical stations. For the well-studied Haut-Ivindo region of Gabon, Brosset (1966) recorded only 24 kinds. In two periods of study at Sengwa Wildlife Research Area in Zimbabwe, Fenton and Bell (1981) detected 23 insectivorous species, and in that region perhaps an additional eight kinds of frugivorous pteropids are possible, for a syntopic bat community of 31 species. Rautenbach, Fenton and Braak (1985) list 41 species for Kruger National Park in northern South Africa. The only places in the Australasian region where neotropical communities could possibly be approached in richness are the lands of the Sunda Shelf and mainland southeast Asia, but Francis (1990) took only 33 species at ground level in the Sepilok Forest Reserve, Sabah.

The Neotropical region is not only the leader in local species richness, it leads in the taxonomic distinctiveness of its fauna as well. Of the nine bat families which occur there, six are endemic, as are 81% of its genera.

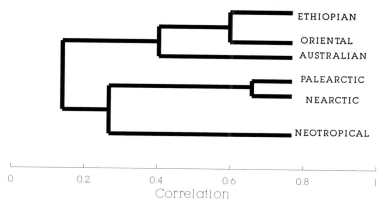

Fig. 6.4. Relationships of bat faunas of the classical zoogeographic regions. The diagram is an UPGMA phenogram (Sneath and Sokal, 1973) derived from correlations between the regions based upon their shared and exclusively held genera. (Data from Koopman, 1970.)

The Old World tropics cannot be clearly divided into the classical zoogeographic regions on the basis of bat distribution. Eight paleotropical bat families range across the region, with one (Rhinopomatidae) dropping out east of Wallace's Line. One distinctive family is based on a single species (*Craseonycteris thonglongyai*) known only from Thailand. Many genera range throughout the Paleotropical region. The Australian region is not delineated by the occurrence of endemic families as it is in the case of the other mammals and of birds, although it does support three endemic subtribes of megabats and one of vespertilionids. The Nearctic region is the world's taxonomic 'no-bat-land' with all of its three families and 80% of its 16 genera shared with adjacent regions. When the bat faunas of each of the classical zoogeographic regions are compared, based upon the genera they have in common and upon the number of species in these genera that occur in each region, it is seen that there are two major regions: the Neotropics and the Paleotropics. The Nearctic and Palearctic regions are closely similar, and related to the Neotropics through the vague resemblance of Nearctic and Neotropical regions. This Holarctic bat assemblage is characterized chiefly by its depauperate nature and its fauna of hibernating vespertilionids (Fig. 6.4).

Why are bats so speciose in the tropics? That is part of a larger question which has received extensive coverage since the first tropical naturalists began publishing. Stevens (1989) tallied 12 explanations for the latitudinal species richness gradient in general. J. W. Wilson (1974) discussed factors that apply especially to bats and emphasized the often made point that additional kinds of resources are available in the tropics, and that all resources are available for much or all of the year. In the neotropics, fruit, nectar, pollen, small vertebrates, blood, fish, and arthropods of all sizes are fed upon by bats, but it has been asserted (Janzen and Wilson, 1983) that fruit and flowers account for the greatest percentage of the increased diversity. Paleotropical bats are less diverse in their feeding habits (no vampires, possibly no piscivores). An important feature of the forested tropics that is especially important for bats, beyond the simple diversity of dietary items, is the fact that the environment is chiefly three-dimensional. Certainly bats, as flying animals, live in a three-dimensional world wherever they occur, but in vast reaches of the temperate zones, the airspace through which bats forage is relatively uncluttered by vegetation or other obstacles. The chiefly insectivorous bats (mostly vespertilionids) of these regions are essentially limited to the aerial pursuit of flying insects. In tropical forested regions, in addition to uncluttered airspace, 30 m or more of complex habitat exist above ground level. This habitat provides extensive surfaces of vegetative substrata whereon a diversity of invertebrates and small vertebrates may be perched, to be picked off by the appropriately specialized kind of bat. Pursuit in the open air may take place over the canopy, in clearings, or along watercourses. Thus, in tropical forest regions, a much greater array of opportunities for insectivorous bats exists than is the case in most parts of the temperate zone, and a wide diversity of insect-feeding species is present. It is not only frugivorous bats that produce the tropical species richness, however, as a simple tabulation of insectivorous kinds will quickly show. Availability of spatial and dietary resources probably does not explain richness of forest bats in Africa, where species density is very low, about on a par with northern Mexico or temperate South America.

The question 'why are there so many species?' is two questions: (1) How can so many species coexist? and (2) how did so many species come into existence? The first of these asks how a given set of resources can support a certain number of species, and is an ecological question involving methods of resource exploitation and sharing. The second question asks about the causes of the production of so many species in one area as compared with another. This is a historical–zoogeographical question and involves consideration of the opportunities for speciation and the history of habitats and barriers to dispersal. The questions focus attention on two possibilities: a region can be speciose because many opportunities for making a living exist there, and/or because conditions in the region have been propitious for speciation. The first possibility can be assessed in part by examining the degree to which the species of the region have specialized in a diversity of modes of life. Patterns in this kind of diversity are a chief focus of this chapter. Chapter 7 deals with the second possibility: that regions differ in the historical and geographical opportunities they have provided for speciation.

Patterns in taxonomic and trophic diversity: what the bats are doing there

The point is often made that a fauna composed of 20 species belonging to the same genus is less diverse than one composed of 20 species divided among a number of genera and families. This is so because the taxonomic decision to allocate species to different higher categories is based upon the recognition that some species resemble each other more than they do others, that natural morphological groups of species occur. These morphologically defined groups traditionally form the basis for erecting genera, families, and higher categories. Of course it is increasingly recognized that these higher categories reflect genetic differentiation and diversity. The morphological features of a species are a guide to its ecological functioning, with the result that higher taxonomic categories can usually be interpreted as ecological categories as well. For this reason, if the 20 species are divided into four genera this probably means that four major modes of life are represented, and that such a community is more diverse ecologically than the community where all 20 species belong to the same genus. For example, a local montane bat community in New Mexico might consist of three species of *Myotis*: *M. volans*, *M. evotis*, and *M. leibii*. A lowland community in the same region might consist of *Myotis thysanodes*, *Pipistrellus hesperus*, and *Antrozous pallidus*. Although each community has the same diversity if merely the number of species is considered, the ecological diversity of the two is markedly different. All three montane *Myotis* are aerial insectivores, although there are differences in the amount of time each spends in various sorts of foraging behaviors. In the lowland community, the *Myotis* is an aerial insectivore/gleaner of soft-bodied insects, the pipistrelle is an aerial pursuer of small, probably mostly harder-bodied prey, while the *Antrozous* is a gleaner and terrestrial predator of larger invertebrates and even small vertebrates. The lowland community is clearly more diverse ecologically, but a simple enumeration of species does not reveal this. However a count of *genera* would show the difference. Thus, expressing

Table 6.1 *Generic diversity of bats in the classical zoogeographic regions*

H', Shannon's index, is a measure of the uncertainty involved in the genus to which a randomly drawn species belongs. Higher values indicate greater uncertainty. Hill's number, $N2$, is the inverse of Simpson's index. If species were distributed uniformly among genera it would equal the number of genera in the region; lower values result from lower evenness. It may be interpreted as the number of very speciose genera in the region. J' measures evenness of distribution of species among genera, and represents H' as a fraction of the value H' would take if all genera contained an equal number of species. A value of 1.0 would indicate equal numbers of species in all genera. See Ludwig and Reynolds (1988) for further discussion

Region	Genera (species)	H'	$N2$	J'
Neotropical	67 (221)	3.88	44.4	0.92
Nearctic	16 (40)	2.35	8.0	0.74
Palearctic	23 (85 +)	2.66	10.6	0.72
Ethiopian	44 (186)	3.33	23.6	0.84
Oriental	57 (268)	3.28	15.7	0.57
Australian	48 (166)	3.36	19.1	0.65

diversity at different taxonomic levels may provide different ecological insights into community dynamics.

Accordingly, diversity at the generic and at the familial levels for bats in the classical zoogeographic regions is summarized in Tables 6.1 and 6.2. In these tables the number of genera and of families per region are shown, and the number of species in each genus and in each family in each region are used to calculate certain other diversity indices as follows: (1) Shannon's information statistic H', (2) Hill's number $N2$, which is the inverse of the Simpson index (Λ), and which represents the number of very speciose genera or families in the community, and (3) Pielou's J', a measure of the evenness with which species are distributed among genera or families. These measures, discussed by Ludwig and Reynolds (1988), support several generalizations concerning taxonomic diversity.

At the generic level, the Neotropical region is outstandingly the most diverse and has the most even distribution of species among genera. Paleotropical areas are less diverse and have a less even distribution of species. In the paleotropics, the genera *Pteropus, Rhinolophus, Hipposideros, Myotis* and *Pipistrellus* contain large numbers of species and contribute heavily to the difference in evenness. Those genera are most speciose in the Oriental or Australian tropics. All have undergone recent bursts of

Table 6.2 *Familial diversity of bats in the classical zoogeographic regions. Symbols as in Table 6.1*

Region	Families	H'	N2	J'
Neotropical	9	1.35	2.7	0.60
Nearctic	3	0.77	1.8	0.69
Palearctic	8	1.03	1.8	0.47
Ethiopian	8	1.66	4.4	0.80
Oriental	9	1.46	3.5	0.75
Australian	6	1.41	3.5	0.80

speciation aided by the insular nature of the region, and as a consequence are represented by many locally distributed endemic species.

At the familial level, the Paleotropical region is the most diverse and also shows the greatest evenness in distribution of species among families. In contrast, a single family, the Phyllostomidae, accounts for the great specific and generic diversity of the neotropics.

Thus, at the generic level the greater diversity is seen in the neotropics, whilst at the familial level it is seen in the paleotropics. Which measure more accurately reflects the amount of ecological diversification in the two faunas?

Wilson (1973) allocated 169 genera of bats (all those recognized at the time) to seven trophic categories:

1. Carnivores (feeding upon tetrapods).
2. Piscivores (feeding upon fish).
3. Sanguinivores (feeding upon blood).
4. Foliage gleaners (capturing insects on foliage or on the ground).
5. Aerial insectivores (capturing insects in the air).
6. Frugivores (eating fruit).
7. Nectarivores (feeding on flowers, nectar, and pollen).

Each genus might feed exclusively in one category or fractionally in several categories. The importance of each genus in each zoogeographic region was calculated, based upon the proportional representation of its species among the total species of the region. For example, a genus with 10 species in a region supporting 100 species would have an importance index for that region of 0.10. For each region, each generic importance index was multiplied by the trophic value(s) of the genus, and the products summed across all the genera in the region to provide an index

Table 6.3 *Relative importance of seven trophic modes (1–7) in each zoogeographic region after Wilson (1973)*

Region	1	2	3	4	5	6	7	N2	J'
Neotropical	1.8	9	1.4	10.1	43.6	30.2	13.3	3.9	0.80
Nearctic	0	3.7	2.4	13.6	75.4	1.0	3.8	1.7	0.51
Palearctic	0	2.6	0	12.0	84.0	0.7	0.7	1.4	0.52
Ethiopian	0.3	0.3	0	14.3	65.8	13.8	5.5	2.1	0.63
Oriental	0.6	0.9	0	15.6	60.5	19.5	3.5	2.4	0.69
Australian	1.0	0.3	0	9.3	48.4	35.1	6.4	2.8	0.79

Notes:
1, carnivory; 2, piscivory; 3, sanguinivory; 4, foliage gleaning; 5, aerial insectivory; 6, frugivory; 7, nectarivory; N2 would equal 7 if each trophic mode were equally important in a region, and may be interpreted as the number of very important modes.

to the importance of each trophic activity in each region. These results are summarized in Table 6.3. Most bats in all six regions are aerial insectivores. Only in the Neotropical and Australian regions is frugivory important. Other categories are relatively unimportant everywhere. The Neotropical region is seen to be the only one with representatives in each category and seems to be the most diverse trophically. This appearance is supported if the trophic importance values for each region are used to generate diversity and evenness indices, which are included in Table 6.3. Based on these indices the Neotropical region is the most diverse, the two temperate regions least, and the three Paleotropical regions are intermediate and similar to one another, not surprisingly about the same result seen in the generic diversity analysis. The Neotropical region also has the most even distribution of species among trophic categories, and is closely followed by the Australian region. Generic diversity provides a better index of the trophic diversity categories of Wilson than does familial diversity.

The assumption of this analysis is that there are only seven feeding categories and that all bats in a given category are equivalent. For example the small, lightly wing-loaded, low aspect-ratio neotropical *Thyroptera* together with the large, heavily wing-loaded, high aspect-ratio, neotropical *Eumops* are both categorized in Wilson's system as aerial insectivores. While it may be true that both capture insects on the wing, to group the two trophically is like grouping cougars and ermines because both capture warm-blooded prey on the ground. Obviously a great deal of ecological diversity is obscured by the coarseness of the trophic resolution. One has to place animals in feeding categories according to the best information to hand, however, and for most bats details of food intake and of foraging behavior are simply not available. But the naturalist *knows* that *Eumops* and *Thyroptera* have diets that are nearly mutually exclusive, and that their foraging methods are totally different. How do we know that? Because, for one thing, the bats are built very differently, and the correlation between form and foraging–feeding behavior in animals is too well established to allow us to ignore it. How then may we take this knowledge, based upon structural differences, and quantify it so that more refined calculations of diversity may be accomplished? One suggestion comes to us from the field of ecomorphology, the study of the relationship between morphology and ecological functioning.

Patterns in morphological diversity: how the bats are built, and what that tells us about their ecology

Ecomorphology builds upon the relationship between form and function to investigate the ways in which detailed morphological differences translate into ecological differences. Armed with this information, the ecomorphologist hopes to be able to gain insight into the ecological relationships between coexisting species. Conventional field investigations into these questions are time consuming and yield information slowly, as we have seen. Perhaps analysis of a set of morphological measurements, easily obtained from museum specimens, or even gleaned from the literature, can provide us with some short cuts to understanding, and help us to focus our field studies in a more efficient way.

To illustrate the use of multivariate techniques in quantifying morphological and trophic diversity, and to supplement the understanding of chiropteran diversity gained from taxonomy and trophic categories, we shall examine assemblages of bats from Italy, New Mexico, Ghana, West Malaysia (the Malay Peninsula), and the territory of Amazonas, southern Venezuela. These five places represent major zoogeographic regions, do not differ greatly in area, and have been the subjects of reasonably comprehensive recent faunal studies (Table 6.4). Measurements of eight external features taken from preserved museum specimens serve as a sample of the morphology of the species. The eight

Table 6.4 *Description of places from which faunal samples have been analysed with respect to morphological variability*

Political entity	Zoogeographic region	Total species	Species sampled	Area (km^2)	Latitudinal limits	Reference
New Mexico, USA	Nearctic	25	25	315 115	37°N–32°N	Findley et al. (1976)
Italy	Palearctic	27	27	251 730	46°N–36°N	Toschi and Lanza (1959)
Amazonas, Ven.	Neotropical	98	86	175 750	6°N–1°N	Handley (1976)
Ghana	Ethiopian	52	43	238 539	11°N–5°N	Rosevear (1965)
West Malaysia	Oriental	80[a]	59	132 090	7°N–2°N	Medway (1969); Koopman (1989)

Notes:
Ven., Venezuela.
[a] 20 species are insular or known only from one or two specimens.

attributes (head and body length, tail length, tibia length, forearm length, ear length, length of fifth metacarpal and phalanges, length of wing tip) are related to the way in which the bats forage and feed. Aerial pursuers and species that forage over large areas tend to have long narrow wings and smaller, laterally directed ears. Hovering gleaners and 'fly-catchers' (species that hang from a perch from which they sally forth to capture prey) tend to have shorter, broader wings, and large, forwardly directed ears. Aerial insectivores often have extensive tail membranes and long tails, because the tail membrane, as well as the wings, may be used in bagging flying insects. Species that capture larger prey, such as small vertebrates and terrestrial arthropods, by mouth have reduced tail membranes, and most frugivores lack tails entirely. From a practical standpoint, these measurements are selected because they are ones that are easily taken by field workers at the time the animals are captured, or because they can be accurately taken from preserved museum specimens. Other measurements could be used, and indeed many measurements of cranial and dental features are routinely studied by bat systematists and ecologists, and many of these have clear relevance to trophic activity. Tooth structure, for example, clearly points to the kinds of food which the bat is best equipped to handle. The set of external measurements used here, however, serves as a good exemplar of the morphological-diversity patterns that bats display on a world-wide basis. Inclusion of additional traits would provide increased depth and refinement to our understanding.

Morphological variability

Comparison of eigenvalues from the analyses (Table 6.5) shows that the tropical faunas are much more variable than their temperate counterparts. This is to be expected simply because there are more kinds of tropical bats. Amazonas is seen to be more variable than Ghana and Malaya, as trophic and generic diversity figures suggest it should be (Tables 6.1, 6.3). Morphological variability may also be assessed by computing the coefficients of variation for each measurement in each fauna, and then using the mean of these coefficients as an index to total variability. Results of these calculations are also tabulated in Table 6.5. Here the same pattern is seen as in the comparison of eigenvalues. Table 6.6 contains correlations between the various diversity measures calculated for the five bat faunas. All the correlations are in the expected positive direction, although only three are statistically significant when

Table 6.5 *Measures of morphological variability for bats from five faunas*

The eigenvalues come from a principal component analysis of log-transformed measurements. Larger sums indicate greater morphological variability. The coefficients of variation (CV) are those of eight morphological variables measured for all bats in each fauna.

Location	Summed eigenvalues	Mean CV
New Mexico, USA	0.0842	0.075
Italy	0.0898	0.061
Amazonas, Venezuela	1.3354	0.331
Ghana	1.2470	0.238
Malaya	1.0060	0.199

Table 6.6 *Correlations between taxonomic and morphological measures of diversity and trophic diversity*

Trophic diversity, an ecological measure, is best predicted by species richness.

	E	T	F	CV	S	G
Eigenvalue[a] (E)	1					
Trophic diversity (T)	0.77	1				
Familial diversity[a] (F)	0.81	0.25	1			
Coefficient of variation[a] (CV)	0.96★	0.90	0.62	1		
Number of species[b] (S)	0.82	0.99★★	0.33	0.94	1	
Generic diversity[b] (G)	0.80	0.94	0.33	0.94	0.96★	1

Notes:
★★$P=0.01$.
★$P=0.05$.
[a]morphological.
[b]taxonomic.

the Bonferroni adjustment for multiple comparisons is applied. None-theless it is clear that (a) morphological diversity indices are reflective of trophic diversity, (b) the mean coefficient of variation is much better in this regard than is the eigenvalue, (c) family diversity is not strongly related to any of the other measures, and (d) best of all for predicting trophic diversity is species richness. Analysis of more extensive sets of measurements would improve substantially the ability of measures of morphological diversity to predict trophic diversity. However, the species ideally provides a summary of the morphological attributes of each kind of bat. It emerges as the single best index to trophic diversity.

Table 6.7 *Loadings of eight morphological characters on three principal components*

Larger absolute values indicate greater involvement of the character with the given component. All characters except 'tail' are involved with PC 1, but 'ear' is somewhat less important than the others. Bats high on PC 1 are small. PC 2 is a tail–ear variable. Bats high on this component have very short or absent tails and somewhat short ears. PC 3 is an ear–tail variable. Bats high on this component have long ears and short tails.

Character	Component		
	1	2	3
Ear	−0.620	−0.431	0.650
Tip	−0.973	0.029	−0.098
Tibia	−0.953	−0.083	−0.013
Foot	−0.956	0.088	−0.057
Tail	0.166	−0.932	−0.319
Head and body length	−0.938	0.080	−0.202
Forearm	−0.976	−0.042	−0.117
Digit 5	−0.980	0.047	0.010

Functional interpretation of principal components

The contribution of each measurement to each principal component can be determined from the factor-loading pattern (Table 6.7). Large negative or positive values for a measurement indicate that it makes an important contribution to the principal component, while values closer to zero indicate little involvement. For example, principal component 3 (PC 3) for New Mexico expresses ear and tail length. Examination of the New Mexican species arrayed on PCs 2 and 3 shows the huge-eared, relatively short-tailed plecotines *Euderma*, *Idionycteris*, and *Plecotus* occupying the high end of the PC 3 axis, while the short-eared, long-tailed *Lasiurus* is found at the low end. Morphograms for each fauna are depicted in Figs 6.5 through 6.9 and one for all five faunas combined is shown in Fig. 6.10.

In each fauna, PC 1 is a size factor. When measurements of dimensions are analyzed, absolute size almost always turns out to be the most variable feature, and every dimension is correlated with size. Thus the large correlations or 'loadings' of all the variables on one factor means that large bats score at one extreme and small bats at the other. Note that tail length is not involved in PC 1. The very largest bat, *Pteropus*

New Mexico

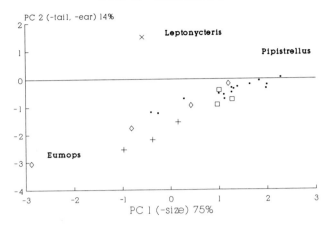

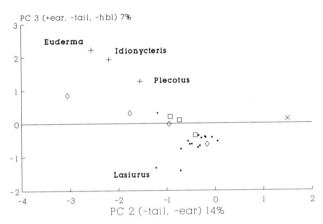

Fig. 6.5. Morphograms of New Mexican bats arrayed on three principal components. The zero-intersection marks the average bat of all the communities considered in this section (Italy, Amazonas, Ghana, Malaya, New Mexico). Thus a majority of New Mexican species are longer-tailed and longer-eared than average. These are traits of aerial insectivores, many of which utilize the tail membrane in insect capture. New Mexican bats are also smaller than the world average, as are most strictly aerial insectivores. Larger species are longer-eared and longer-tailed: plecotines and the large molossid *Eumops perotis*. PC 3 separates large-eared plecotines from other vespertilionids which practice aerial pursuit more than gleaning. (■ other vespers; + plecotines; □ gleaning *Myotis*; × *Leptonycteris*; ◇ molossids)

Italy

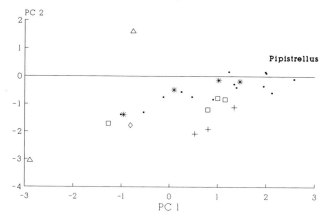

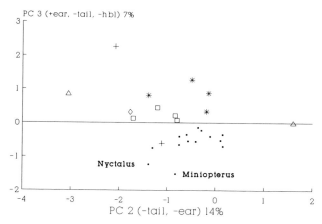

Fig. 6.6. Morphograms of Italian bats. A chief difference from the New
Mexican fauna is in the addition of *Rhinolophus* which occupies a sector of
morphospace that is vacant in North America. Italian vespertilionids appear to be
more dispersed in morphospace than do their relatives in the New World.
(■ other vespers; + plecotines; ★ *Rhinolophus*; □ gleaning *Myotis*; ◇ *Tadarida*;
△ outliers)

Amazonas

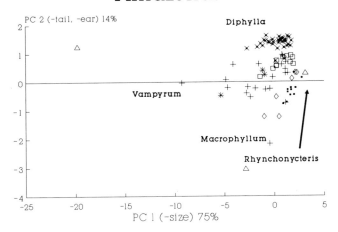

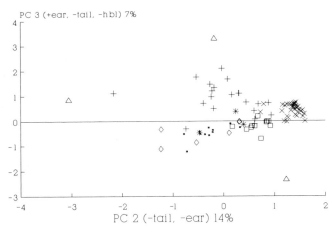

Fig. 6.7. Morphograms of bats from Amazonas Territory, Venezuela. This fauna occupies more morphological space than the temperate ones occupy. A principal difference is the presence of a large number of frugivorous species (including the flower-visiting glossophagines) which are positioned high on PC 2, indicating short or absent tails and short ears. Gleaning species, mostly phyllostomines, form a distinct cluster positioned approximately as are plecotines in the temperate faunas, except that, because phyllostomines are often short-tailed, they appear higher on PC 2. Vespertilionids are rare in this fauna. The emballonurids are commoner aerial insectivores, but because they are relatively shorter-tailed than vespertilionids, they, too, appear higher on PC 2. (■ vespers; + phyllostomines; ★ Noctil-Ptero; □ emballonurids; × frugivores; ◇ molossids; △ outliers)

Ghana

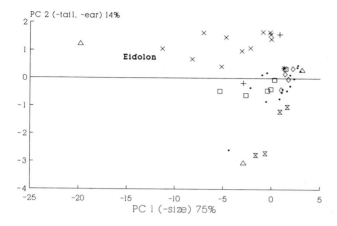

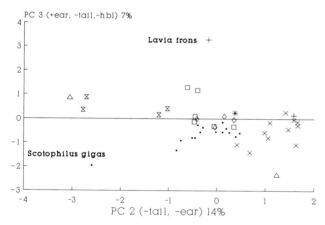

Fig. 6.8. Morphograms of the Ghanaian fauna. Compared with the Amazonas fauna, this one contains fewer species, but they are more widely dispersed in morphological space. The frugivores occupy a similar position to New World frugivores, but they are much more widely dispersed, and some are much larger in size. Rhinolophoid bats occupy approximately the same space as New World gleaners. (■ vespers; + megaderms; ★ *Rhinolophus*; □ *Hipposideros*; × pteropids; ◇ molossids; △ outliers; x *Nycteris*)

Malaya

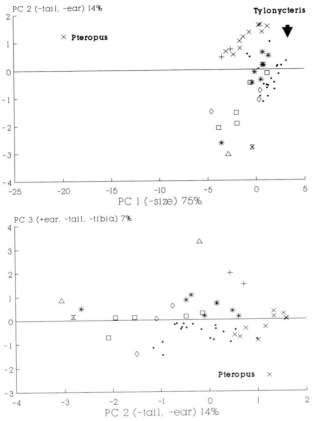

Fig. 6.9. Morphograms of the fauna from the Malay Peninsula. Compared with the African fauna, the Malayan one is less dispersed in morphospace, especially with respect to the pteropids, which are generally smaller. (■ vesperoids; + *Megaderma*; ★ *Rhinolophus*; □ *Hipposideros*; × pteropids; ◇ molossids; △ outliers; x *Nycteris*)

vampyrum (Fig. 6.9), has no tail, nor does the largest neotropical bat, *Vampyrum* (Fig. 6.7). The smallest bat (largest PC 1 score) is the Asian vespertilionid *Tylonycteris pachypus*. The largest is *Pteropus vampyrum*. These species are used as outliers on PC 1.

The second component, PC 2, involves tail and ear length. Bats placed on the positive end of PC 2 have short tails and short ears. Temperate-zone faunas lack bats at this extreme, except for the flower bat *Leptonyc-teris* in New Mexico. All of the frugivorous species, pteropids in the Old

World bat morphology

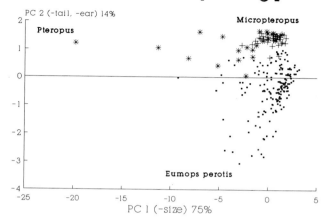

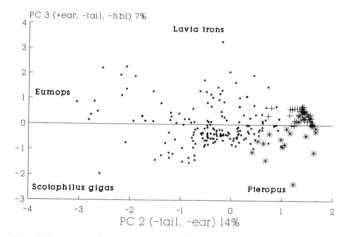

Fig. 6.10. Morphograms of the five faunas combined. Note that the distribution of points in morphospace is bimodal, with frugivores forming one, and insectivores the other, mode. Extreme bats on each component are identified, except for the small extreme on PC 1 which is the Malayan *Tylonycteris pachypus*. As a group, frugivores are less variable morphologically than insectivores and carnivores. (+ NW frugivores; * pteropids)

World and phyllostomids in the New, are tail-less or possess only very short tails. These kinds are grouped high on PC 2 in all the tropical faunas. The division of bats into tail-less frugivores and tailed insectivores is the most basic ecomorphological division on a world-wide basis. The tail is often of importance in insect capture, and also in slower, more controlled flight. For frugivores and large carnivores, an extensive tail membrane is an aerodynamic impediment. Species low on PC 2, plecotines, *Nycteris*, and some rhinolophids, for example, are outstanding in their ability to hover and glean, using slow, maneuverable flight. The shortest-eared tail-less bat is the African pteropid *Micropteropus*. The longest-eared longest-tailed bat is the New World molossid *Eumops perotis*. These are the outlying species on PC 2.

The third component, PC 3, combines ear, tail, and to a lesser extent head and body length. Bats high on this factor are large-eared and relatively short-tailed, while at the opposite extreme are species with long tails and short ears. In the former category are megadermatids (large-eared carnivorous species lacking tails), phyllostomines, the large-eared gleaners of the neotropics, and plecotines. In the latter are most vespertilionids, and other aerial pursuers. The longest-eared shortest-tailed bat is the African megadermatid *Lavia frons*. The relatively shortest ears are seen on *Pteropus*, and even though it is tail-less, it appears low on PC 3. These two are outliers on PC 3.

In this analysis, using eight characters, there are five or six significant morphological–ecological axes, with bats arranged in continua, or sometimes in clusters, along each. Bats on the extreme of each axis represent different foraging or trophic categories; the limited dataset used in this example reveals 10 or 12 such groupings. If we added data on other obvious external features, such as nose-leaves, the groups would multiply. And if internal features such as teeth and jaw structure were included, still finer subdivision would be achieved. There seem to be more trophic–foraging categories in each fauna than the seven used in Wilson's analysis of the biogeographic regions, but the eight measurements used do not reveal the total spectrum of variability.

African bats: two examples of ecomorphological correspondence

Although the temptation to rely upon the correspondence between morphology and ecology is beguiling, rather few convincing demonstrations of the relationship exist. One examination of the actual relationship between diet and morphology in bats is that of Findley and

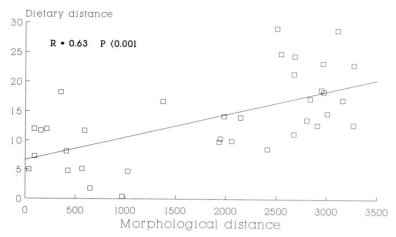

Fig. 6.11. Morphological differences between all pairs of nine species of Zambian bats compared with dietary differences between the same species pairs. The more two species differ morphologically, the more they differ in diet (after Findley and Wilson, 1982).

Black (1983). Black made collections of insectivorous bats in the Lusaka area of Zambia, East Africa, over a period of one and a half years. The stomachs of all the bats were analyzed by Whitaker (Whitaker and Black, 1976), and the bats themselves were preserved as museum specimens. Subsequently 11 external measurements were recorded for samples of the nine species in the collection, and the dietary information was grouped under 19 headings. Comparisons of the morphological and dietary datasets were presented by Findley and Black (1983) and by Findley and Wilson (1982). Findings included the following:

1. Morphological distances between pairs of species were positively correlated with dietary differences between the same pairs of species: the greater the morphological difference the greater the dietary difference (Fig.6.11).
2. Morphological differences between pairs of species were inversely correlated with the degree of dietary overlap between the same pair of species: the more two species resemble each other morphologically the more their diets overlap (Fig. 6.12).
3. Morphological variability of each species was positively correlated with its dietary variability: the more variable the morphology of a species the more variable its diet (Fig. 6.13).

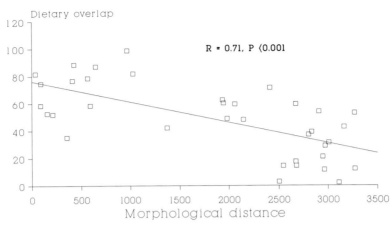

Fig. 6.12. Morphological difference between all pairs of nine species of Zambian bats compared with dietary overlap. The more two species resemble one another in morphology, the more their diets overlap (after Findley and Wilson, 1982).

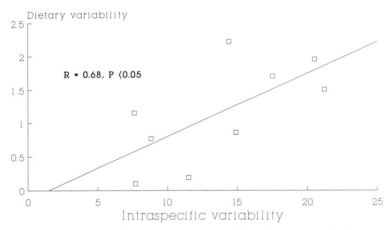

Fig. 6.13. Intraspecific morphological variability compared with dietary variability of each of nine species of Zambian bats. The more variable a species in morphology, the more variable its diet (after Findley and Wilson, 1982).

To relate the data from these studies more directly to those discussed in this chapter, the same eight external measurements were used, and a principal component analysis (PCA) performed to extract scores for each Zambian bat on three morphological components. The dietary data were likewise submitted to a PCA analysis, and scores of bats on the first

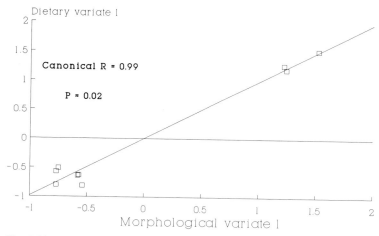

Fig. 6.14. Comparison of morphological and dietary attributes of nine species of insectivorous Zambian bats (based upon a canonical correlation analysis of data from Findley and Black, 1983). Morphological and dietary attributes are highly correlated.

three dietary components were computed. Then the three new morphological scores and the three new dietary scores for each of the nine kinds of bats were compared using canonical correlation analysis. The result is three new morphological and three new dietary variables, each called a canonical variable, so constructed that morphological variable 1 is maximally correlated with dietary variable 1 within the limits imposed by the data, morphological variable 2 is maximally correlated with dietary variable 2, and so on. In the case of the Zambian bats, the results are shown in Fig. 6.14. Morphology and diet are highly correlated, and most of the variability in the two datasets is expressed in the relationship pictured.

Findley and Black (1983) summarized their conception of the structure of the community they studied in a diagram (Figure 6.15) which suggests a group of similar, invariable species overlapping substantially in diet, and fewer, dissimilar, more variable species which overlap less with their neighbors.

In a second ecomorphological comparison, Aldridge and Rautenbach (1987) worked with 26 species of microchiropterans in Kruger National Park, South Africa. For each they measured morphological attributes affecting flight (wing loading, aspect ratio, and wing-tip shape), attributes of the echolocation cries (frequency, intensity, duration, and

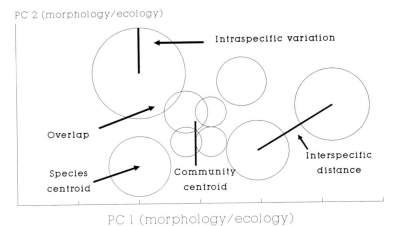

PC 2 (morphology/ecology)

Intraspecific variation

Overlap

Species
centroid

Community
centroid

Interspecific
distance

PC 1 (morphology/ecology)

Fig. 6.15. A bat community in attribute space. The attributes may be morphological or ecological. The community consists of a core of similar, invariable, overlapping species, and a periphery of more distinctive, variable kinds, which overlap little with their neighbors (after Findley and Wilson, 1982).

shape), maneuverability, habitat use, and prey size. The investigators hypothesized that (1) wing morphology and echolocation-call design can determine foraging site selection and foraging behavior, and (2) that echolocation-call design should be compatible with wing morphology. In support of the first hypothesis, significant correlations between the listed variables were demonstrated. In support of the second, significant correlations were reported between those morphological traits that improve maneuverability and echolocation calls that are resistant to acoustic clutter. Thus, as in the Zambian study, there is good ecomorphological correspondence. Aldridge and Rautenbach also showed a relationship between foraging habitat and diet, and between the sizes of predator and prey. Finally, large bats were seen to take a variety of sizes of insects, while small ones took only small prey.

In summary, it is possible to have a reasonable amount of confidence in the ability of morphology to provide an insight into the feeding and foraging of insectivorous bats.

Patterns in packing: how full is ecomorphological space?

When the bats from each fauna are arrayed in eight-dimensional space, and the distance of each bat from the position of an average bat for that

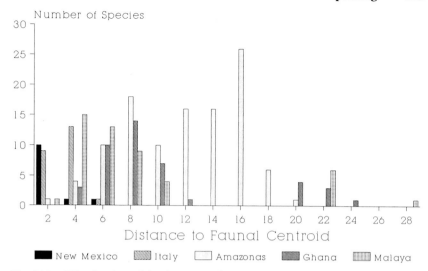

Fig. 6.16. Distribution of the distances of each bat in each fauna from its faunal
origin, the mean position in morphospace of all the bats in the fauna. Based upon
log-transformed data.

fauna, the faunal centroid, is calculated, the dispersal of the species in
morphospace is readily visualized. Figure 6.16 depicts the distributions
of these distances for the five faunas. The faunas of Italy and New
Mexico are closely clumped near the faunal centroid. Morphological
dispersal has not been great, and all the species, mostly vespertilionids,
resemble one another relatively closely. The three tropical faunas consist
of a grouping of species fairly near the faunal centroid, and another
group, of larger or smaller dimensions, at some distance from that point.
The distant group in each case is made up of the tail-less species, mostly
frugivores, and the less excentric group comprises the tailed species. All
three tropical faunas show greater dispersion in morphospace than the
temperate ones. Among the three, that from Amazonas shows the
greatest average dispersion (the greatest mean distance from its faunal
centroid) and that from Malaya shows the least. The actual extent of
morphospace looks greater in the Old World tropics, but the degree to
which morphospace is filled in, or packed, is much greater in the
neotropics. The morphospace of the paleotropics, especially of Africa,
has large gaps in it; that of the neotropics is more evenly filled. In general,
as is evident in the morphograms presented here and in Chapter 4, each
fauna is composed of many similar, closely packed species, and a

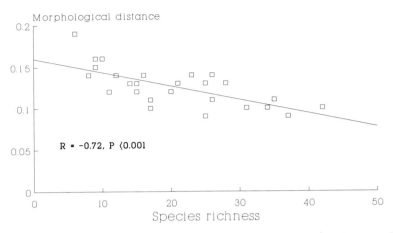

Fig. 6.17. Morphological packing as a function of species richness in New World insectivorous–carnivorous bat communities. Packing increases (distance between nearest neighbors diminishes) as richness increases. The best-fit regression is a power function, $Y = 0.3X^{-0.3}$, suggesting a lower asymptote to packing. Modified from Schum (1984).

scattering of more distant, dissimilar kinds. As noted by Fleming (1986), Schum (1984), and others, the level of packing increases with species richness: richer faunas contain more closely packed, but also some more widely dispersed, species. Schum's study of communities of insectivorous/carnivorous bats over a gradient from Canada to Central America showed that packing in the central core of the community increased with richness in both real and random assemblages (Fig. 6.17). However, since the larger random communities came close to mimicking the larger real ones, the similarity between random and real is not unexpected, and also serves to emphasize that larger bat communities, as Fleming suggested, are truly reflections of the regional faunas of which they are a part. These long-noted packing patterns are not limited to bats, and may be universal among faunal or community assemblages of organisms (Van Valen, 1973; Ricklefs and Travis, 1980; Fleming, 1986).

Patterns in biomass and numerical abundance

What relationship do biomass and abundance bear to the patterns thus far revealed? Difficulties in evaluating numerical abundance and biomass have been described in Chapter 3. Indeed, few chiroptologists have attempted to put together information comparing various regions. One who has is Gaisler (1979), who summarized density information for a number of kinds of, mostly temperate zone, bats. The estimates range from 0.01 to 3.0 individuals per hectare, except for an estimate for the neotropical disk-winged bat *Thyroptera tricolor* of 21.9 per hectare. *Thyroptera* occurs in small groups which spend the day in rolled young leaves of *Heliconia*, a neotropical banana relative. In the cited study (Findley and Wilson, 1974), most individuals were marked, the home ranges of the groups were determined and this, coupled with knowledge of the extent of the *Heliconia* stand, allowed a fairly precise estimate. However, extensive areas of the forest in which the study took place (Puntarenas Province, Costa Rica) lack the appropriate plants, so that the density, measured over the whole forest, would be lower than that reported. The only other tropical species cited by Gaisler is the Guamanian megabat *Pteropus mariannus* at 0.6 individuals per hectare, well within the range of the various Temperate Zone records. But since it is likely that this species, now extinct, was declining at the time it was studied, it cannot be taken as typical of tropical bats.

In a 10-year study of the short-tailed fruit bat *Carollia perspicillata* in Guanacaste Province, Costa Rica, Fleming (1988) estimated that densities were at from 0.06 to 0.07 individuals per hectare if the bats were uniformly distributed over their foraging range. Since the animals are actually clumped, while foraging, in patches of appropriate food plants, the density in the used areas is much greater. Nonetheless the same qualification applies to the density estimates for temperate zone species presented by Gaisler.

Other published records of bat population density include those of Findley and Wilson (1982), in which the net-night, bats captured per net per night, was used to compare densities in several regions. According to their figures, the greatest densities occurred in the United States, 7.6 per net-night (504 net-nights tallied), followed by the neotropics, 3.6 per net-night (1730 net-nights), and finally by tropical Africa, 1.8 per net-night (427 net-nights). These data for the United States were strongly biased by the kind of netting upon which most of the records are based. Especially in the West, bats are best netted over water holes where they congregate in enormous numbers. The bats swarming around a watering place in the arid parts of the country may then disperse over many miles to forage or to return to shelters. Nets set in other places, except for those set over the entrances to the shelters themselves, rarely capture many individuals. By contrast, in the neotropics, bats are commonly netted as they fly along trails through the forest on their way to foraging grounds. The animals are easily captured in such situations, especially the frugivorous phyllostomids that make up the bulk of most neotropical catches. In the African tropics, bats may be netted over watering places, especially in the more arid regions. However, Fenton *et al.* (1977) saw no evidence of drinking by microchiropterans in Rhodesia, and thought that most obtained sufficient water from their food. Although some American desert species may enjoy increased renal efficiency (Carpenter, 1969; Geluso, 1978), most American species have been observed to drink regularly. In the African forests, water-hole or trail netting rarely produce large catches. In part this may be because there are few frugivorous bats in Africa and those that do occur are not echolocators, and the microbats are much more adept at avoiding nets. Thus various special circumstances conspire to make the comparisons of Findley and Wilson somewhat tenuous.

Heideman and Heaney (1989) estimated densities of pteropids on Negros Island in the Philippines at from 0.2 to 3.7 individuals per hectare for six species, with a cumulative density of about 10 bats per hectare. These densities were found to be correlated with abundance of the bats in netting samples, suggesting that netting data may indeed be used to estimate actual population densities, at least of megabats.

The meager evidence currently available suggests that individual tropical bat species may enjoy higher population densities. The fact that there are many more species of bats in the tropics indicates therefore that density of individual bats there, as well as bat biomass, is higher than in the temperate zone.

The same may apply in the comparison of the neotropical and paleotropical faunas. But one hesitates to embrace this conclusion because of the large size of many of the megabats. Fleming (1988) concluded that, because of their generally much more extensive foraging ranges, megabat resources are probably highly dispersed spatio-temporally, and large size is an adaptation to minimize commuting costs. Thus the implication is that megabat numerical density is actually much lower than that of New World fruit bats, and biomass density in the two regions could be about the same. Much more information is obviously needed before firm conclusions in this area will be feasible.

To the extent that it is possible to generalize, we may suspect that both total numerical density and density per species of bat increase toward the equator, and may be greater in the neotropics than in the paleotropics. Biomass relationships follow those of numerical density. Numerical density and biomass may thus be positively correlated with species richness.

Summary

Species richness in bats is highest in the tropics, and within the tropical world, it is greatest in the neotropics. Richness is a good index of ecological, morphological, and taxonomic diversity, as well as to eco-morphological packing.

Although the evidence is sparse, it seems likely that patterns of population density and biomass reflect that of species richness as well.

Thus the pattern of species richness is of over-riding interest, and an understanding of its correlates may take us a long way towards understanding the development of bat communities.

7 · Correlates of bat community patterns

The caveat against inferring processes from patterns is an article of conventional wisdom among community ecologists (Wiens, 1989). Yet in the case of global patterns, such as those discussed in Chapter 6, the likelihood is slight that experimental demonstrations of the processes behind them will be forthcoming anytime soon. In the meantime the patterns often suggest reasonable process sequences, because of correlations with other well-known empirical or theoretical patterns. Investigation of these correlations may sharpen our focus on possible experimental approaches. Two major patterns to be discussed now involve (1) species richness, and (2) species packing.

Species richness

Species richness shows a classic latitudinal gradient in bats, but of special interest is the fact that, within the tropics, it shows a longitudinal gradient as well. The former has been much, but the latter less commonly, discussed. Why is species richness, with all its correlates of trophic and morphological diversity and numerical abundance so much greater in the neotropics than in any other tropical region?

Community level processes are played out in the context of processes operating on larger geographic scales and over evolutionary, as well as ecological, time. The length of time that a community has been in place influences the degree to which community processes have had the opportunity to act. The physiographic and climatic history of the region are important in determining opportunities for speciation, differentiation, and extinction. Larger areas are increasingly likely to be subjected to geologic or climatic phenomena which cause barrier formation, and isolation of populations, and opportunities for the proliferation of a lineage may be partly determined by the prior presence of competing lineages in the region. I will now look into the possibility that the presence of ecologically similar vertebrates, age, area, and history are related in suggestive ways to the longitudinal gradient in chiropteran tropical species richness.

Sometimes the abundance of one group of organisms is inversely correlated with that of another group. That relationship, referred to as complementarity by Schall and Pianka (1978), has suggested to some investigators the process of competitive displacement or exclusion of one group by another. Schall and Pianka caution that complementarity should not be taken as unequivocal evidence for competition. For example, they note that some North American reptiles are negatively correlated with North American birds, but this relationship is probably driven by an independent correlation with precipitation.

Are there other vertebrate groups knowledge of the abundance of which might help in understanding the longitudinal species–richness gradient in bats?

Fenton (1975) studied echolocation in swiftlets (genus *Collocalia*, family Apodidae) in New Guinea. Some species of this crepuscular and partly nocturnal genus of birds practice low-frequency echolocation which enables them to utilize deep caves as roosts, and perhaps to extend their period of foraging somewhat later in the evening. Fenton speculated that the echolocating species may do some nocturnal feeding, and hence overlap the temporal and spatial foraging niche of certain bats, especially molossid bats which, like the swiftlets, forage above the canopy. The echolocating species of swiftlets are found widely in tropical Australasia in a region notable for its depauperate molossid fauna, a seeming example of complementarity at the family level.

In a later study, Fenton and Fleming (1976) reviewed opportunities for interactions between bats and birds on a global scale. They detected some possible examples of complementarity between fish-eating bats and fish-eating owls. More recent data on carnivory in bats seem to have vitiated some of these conclusions.

Findley and Wilson (1983) in attempting to account for the paucity of frugivorous bat species in African forests compared with their richness in the neotropics, considered the possibility that the presence of other frugivorous vertebrates restricted opportunities for bats. However Africa is depauperate not only in frugivorous bats, but in all other arboreal frugivores as well. Between Africa and Neotropica there is no complementarity between frugivorous bats, or all bats, and other major groups of vertebrates.

Thus, although complementarity may exist in certain restricted cases involving bats, it does not in any case seem to account for the longitudinal gradient in tropical species richness.

Community age

Newly established communities might be expected to have properties which are different than those of old ones. In particular, responses to predation, resource availability, and competitive pressure should have progressed further in older communities. A useful example based upon the relationships of birds of the genus *Parus*, tits and chickadees, has been described by Lack (1969). Five or six species of tits may be found breeding in the same woodlands in parts of Europe and Britain. Here the species are well-differentiated in appearance, in foraging habitats, and in diet. In North America six species of *Parus* occur, but in most places only one or two are to be found. The American species are adapted for existence in different habitats, not for coexistence in the same habitat as is the case with the European species. Lack suggests that the American *Parus* have recently arrived on that continent, and that there has not been sufficient time for the ecological differentiation seen in Europe. In addition he speculates that some tit niches in the New World may have been occupied by warblers or vireos when the *Parus* arrived.

A parallel case involving bats of the genus *Myotis* has been described by Findley (1972, 1976). In Europe and in New Mexico eight or nine species occur. Five occur in Venezuela, and four are recorded from West Malaysia. When the mean taxonomic distance to nearest morphological neighbor is computed for each of these communities, the New Mexican and Venezuelan *Myotis* are each seen to be closely packed, 0.23 and 0.24 distance units respectively, while the European and Malaysian species are more dispersed, 0.42 and 0.48 units, respectively. As with the tits, recency of arrival in the New World was invoked as a possible explanation.

Community age is a function of the length of time that the region has been available for occupancy, and the length of time that the taxon under consideration has occupied it. We now consider these factors.

With one exception it appears that the major contemporary tropical regions have been in place in tropical latitudes since the beginning of the Cenozoic, and tropical forest has existed in them since that time. The exception involves the land masses of the Australian Plate, Sahul Land. New Guinea and northern Australia probably did not enter the tropics until about the Miocene time (Audley-Charles *et al.*, 1981), and since that time the tectonically active and mountainous nature of New Guinea has conspired to make it something less than a stable and enduring tropical forest habitat (Walker, 1982).

The earliest known bats, from Eocene times, are from North America and Europe. The earliest records from the tropics or subtropics are Oligocene (Egypt), and late Oligocene–early Miocene (Brazil). The earliest known phyllostomid is from the Miocene of Colombia. No Southeast Asian or Australian records come from before the Pleistocene. Pre-Pleistocene bat fossils are rare, as are vertebrate fossils from the Oriental and Australian regions and from tropical forests in general. In short, the fossil record of bats is inadequate to allow us to generalize as to the duration of bat occupancy of the major faunal regions. On the available evidence, there is no reason to suppose that any one of the tropical regions has been occupied by bats for a substantially longer time than any other. The possible exception is Sahul Land which may have been available to tropical bats no earlier than the Miocene.

In summary, no paleontological or geomorphological evidence given above suggests that bat assemblages in any major tropical region are markedly older than those in any other.

Area

For diverse organisms, species richness (S) is strongly correlated with the area (A) over which the organisms were counted. Larger areas support more species. So common is this pattern that it has been referred to as one of community ecology's few genuine laws (Schoener, 1986). This relationship is clearest when island biotas are examined, and MacArthur and Wilson (1967) formulated an equilibrium theory of island biogeography to account for it. An extensive literature has developed, and numerous exceptions have been discovered. Wiens (1989) discussed ornithological examples. Nonetheless, whatever the mechanisms may be in a given situation, the relationship between S and A is pervasive in ecology. An exception exists among bats, where no correlation between S and A exists on a world-wide basis (Fig. 7.1a). However, because the relationship between latitude and S is so strong in bats, if the world-wide data are divided into tropical and temperate sets, the A/S relationship emerges (Fig. 7.1b). *Within* the tropics and *within* the temperate zones, larger areas do support more species of bats. Tropical regions vary from nearly bat-less deserts to species-rich rain forests. When the focus is

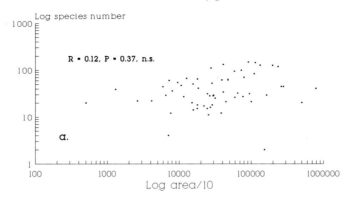

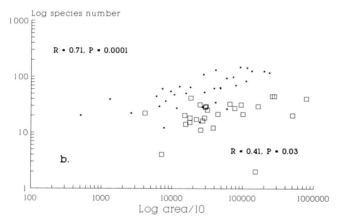

Fig. 7.1. Relationship between area and species richness of bats on a world-wide basis. Within the temperate zones, and within the tropics, species richness increases with area (b), but when all latitudes are considered together there is no significant relationship (a). Each point represents an island or a political unit.

narrowed to *rain forest* area and *rain forest* bats, the S/A relationship is stronger. The New World far exceeds the Old in continuous rain forest area, and it is much richer in bats (Fig. 7.2). Over 80% of the variance in S is accounted for by area.

History

There is strong evidence that during the Pleistocene, tropical forests, and forests generally, were repeatedly fragmented and reunited. During times of fragmentation, forest-dwelling organisms experienced oppor-

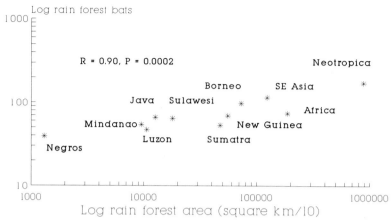

Fig. 7.2. Number of species of forest bats (gamma diversity) as a function of forest area. (Data from various sources.)

tunity for speciation, and it has been suggested that this phenomenon has contributed to tropical species richness (Prance, 1982). Forest fragments constitute refuges for forest organisms, and this body of thought constitutes *Refuge Theory*. Although the tendency to accept refuge theory is widespread, important problems with supporting evidence, and with testing the theory exist (Beven, Connor and Beven, 1984; Lynch, 1988).

If the theory is true in its general outline, it is worth asking if it can help to explain the species richness pattern seen in bats. The more refuges formed in a region the more opportunity for speciation, and the greater the resulting species richness, other things being equal. Estimates of the number of such refuges are available for the neotropics and for Africa, but the situation in the Asian tropics is more difficult to interpret. In South America the number varies from 32 (based upon certain butterflies) to 4 (for *Anolis* lizards), with intermediate estimates of 16 for trees, and 19 for birds (Oren, 1982). For Africa, Grubb (1982) delineates 5 major centers for forest mammals, of which he considers 3 to be the chief sources of immigrants for other regions. However, he notes a total of 13 centers for forest in a wider sense. Laurent (1973) postulated 6 lowland forest refuges for African amphibians and reptiles. A map presented by Haffer (1982) suggests about 27 neotropical refuges, and 6 in Africa. On this evidence, the Neotropical region has provided greater opportunity for speciation by forest organisms during Pleistocene time than has Africa.

In Africa and South America, forest refuges were formed during glacial episodes when cooler, drier conditions resulted in restriction of rain forest to moister enclaves. In the Indo-Australian archipelago the situation is somewhat different. Here the rain forest is currently segmented into numerous refuges on the many islands of the region. During glacial times montane regions were glaciated, and large areas of the Sunda and Sahul shelves became land. These new lands were not necessarily covered with rain forest, however. Since they are of low relief, the likelihood is that much of the newly emerged land supported mangroves or very saline soils. The overal extent of rain forest may actually have been reduced (Walker, 1982) but it is not at all clear that refuge formation would have been any greater, if as great, as at the present time. Nonetheless, the number of forest refuges at the present time is great, at least 10 major ones, and uncounted ones of lesser size. It seems likely that the large number of refuges in the Sunda–Sahul region has contributed to its present biotic richness, even if episodes of speciation were not concurrent with those in Africa or South America.

From the sources cited above, the number of refuges likely to have affected the tropical forest biotas of Africa, Southeast Asia, Borneo, New Guinea, and Neotropica may be estimated. We will use these numbers, together with the rain forest areas of each region, in a stepwise regression analysis with species number as the dependent variable in order to see if the addition of refuge number further explains variance in S. The results are shown in Table 7.1. While area accounts for about 80% of the variation in species number in this analysis, the addition of the number of postulated refuges accounts for an additional 20%. There is a good deal of room for error in the refuge estimates, but if they are at least of the correct order of magnitude the conclusion seems inescapable that history and area come close to being completely explanatory of gamma diversity in tropical bats.

These two factors were also invoked by Karr (1976) to explain diversity differences between the avifaunas of Panama and Liberia. In that case, the forest birds of Liberia were less, and the grassland and savanna birds more diverse than in Panama. Africa supports a much greater area of grassland and savanna habitat and a lesser amount of rain forest than the neotropics, and the parallel between bird and bat diversities in the two places is close.

What of alpha, or single-site, diversity? We must now bring our focus back to the community level. To test the relationship of alpha to area, refuges, and gamma diversity, we will use it in a stepwise regression with those factors as the independent variables. The results are shown in Table

Table 7.1 *Analysis of rain forest area and estimated refuge number as predictors of species number of rain forest bats* (a) *Dataset upon which analysis is based.* (b) *Results of stepwise multiple regression.*

The two independent variables, area and refuge number, account for most of the variation in species number.

(a) Dataset

Region	Area of rain forest (10^3 km^2)	Postulated refuge no.	Species
Africa	1865	6	74
Mainland SE Asia	1230	23	115
Neotropica	8746	27	174
Borneo	725	23	98
New Guinea	554	9	69

(b) Stepwise regression

Step	Variable	Partial R^2	Model R^2	F	P
1	Area	0.7879	0.7879	11.14	0.0445
2	Refuges	0.1997	0.9875	32.03	0.0298

7.2. Only gamma diversity is important in explaining alpha diversity, and it accounts for about 70% of the variability in single-site richness (Fig. 7.3).

Alpha diversity of bats in the African rain forest is less than that in the Neotropical rain forest because there are fewer species of rain forest bats in Africa. And the latter is true because the area of rain forest in Africa is small. The insight of MacArthur and Wilson (1967) illuminates this situation: many of the species of a rich region have access to a single site, thereby enhancing the single-site richness.

Fleming (1986) has shown that neotropical phyllostomid communities closely resemble in taxonomic, trophic, and morphologic structure the structure of the regional phyllostomid fauna as a whole. This finding, in conjunction with the determinants of alpha diversity just discussed, seems to suggest that many community characteristics are to be attributed to factors far removed in space and time from the local situation. Perhaps a Neotropical region containing a rain forest twice as extensive would support local communities of bats with much greater taxonomic,

Table 7.2 *Analysis of rain forest area, postulated refuges, and rain forest species richness (gamma) as predictors of single-site (alpha) species richness* (a) *Data.* (b) *Stepwise regression.*

Only regional richness is significant in accounting for the variance of community richness.

(a) Data

Region	Rain forest area (10^3 km^2)	Number of refuges	Gamma	Station	Alpha
Africa	1865	6	74	Rio Muni	22
				Haute Ivindo, Gabon	24
Southeast Asia	1230	23	115	Pasoh, W. Mal.	25
Borneo	726	23	92	Sepilok, Sabah	33
New Guinea	554	9	69	Central Prov., Papua	21
				Milne Bay, Papua	24
Neotropica	8746	27	174	Boca Mavaca, Ven.	45
				Belen, Ven.	48
				Rio Mavaca, Ven.	40
				San Juan, Ven.	78
				Pto Ayacucho, Ven.	53
				La Selva, CR	65
				Osa Peninsula, CR	50

(b) Stepwise regression

Model: Alpha $= f$ (gamma, area, refuges)					
Step	Variable	Partial R^2	Model R^2	F	P
1	Gamma	0.7033	0.7033	26.1	0.0003

No other variable significantly reduced the variance of alpha.

Note:
W. Mal., West Malaysia; Prov., Province; Ven., Venezuela; CR, Costa Rica.

trophic, and morphological diversity. Rich as they are, neotropical bat communities may not represent the ultimate possibilites.

In summary, gamma diversity of tropical forest bats is explained first by area of forest habitat in the several regions, and secondly by historical opportunities for speciation. Alpha diversity is mostly a function of gamma diversity. Species–rich regions support species–rich communities.

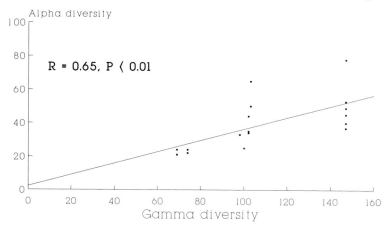

Fig. 7.3. Alpha diversity as a function of gamma diversity. The more species of bat found in a region, the more species are likely to be captured at a single site.

Packing

While community packing per se is correlated with species richness, the form of the pattern is not. All communities, large and small, show the core of densely packed species surrounded by fewer, more distant kinds. One is inclined to suspect the influence of some well-known mathematical regularity. This suspicion is strengthened when it is seen that the distribution of values describing a diversity of phenomena show a similar pattern (Fig. 7.4). Indeed, raw numbers of this kind are often log-normally distributed, with a preponderance of low and fewer high values. Log–normal distributions result when the magnitude of the variables is determined by the multiplicative interplay of many more or less independent factors (May, 1981). Log transformation of variables may result in a Gaussian distribution, but that continues to depict a central cluster of similar values and lesser numbers of excentric ones (Fig. 7.5). The familiar packing pattern may thus very well result from the action of diverse unrelated factors upon the events of speciation and differentiation which lead to the placement of species in ecomorphological space. Is this community pattern thus a statistical artifact of no ecological meaning? The answer is clearly no. Even if in origin that pattern is generated by random processes, the bats still must live with the result. There really *are* a lot of similar invariable bats which overlap one another, and fewer more distant, distinctive, and variable kinds. That is a description of ecomorphological reality however it came about.

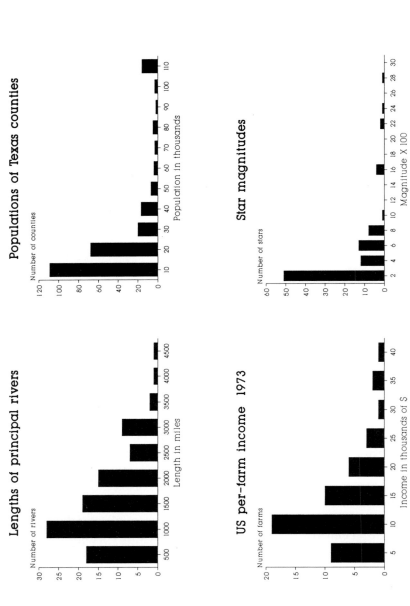

Fig. 7.4. Right-skewed distributions of four variables. The resemblance of these patterns to log-normal distributions suggests that they are determined stochastically.

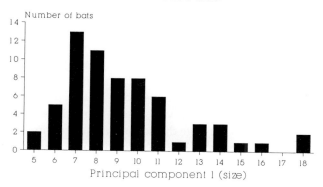

San Juan Bats on PC 1
Standardized data

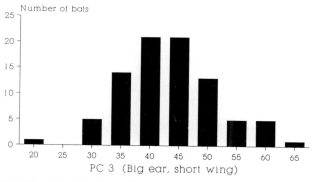

Amazonas Bats on PC 3
Log-transformed data

Fig. 7.5. Distributions of principal component scores for bats from southern Venezuela. The resemblance of these patterns to log–normal and Gaussian distributions suggests that they were formed by stochastic processes.

If biotic interactions between newly speciated congeners lead to optimal ecomorphological spacing, it is likely that the result would not be discernible when packing of the entire community is examined. Accordingly, Willig and Moulton (1989) analyzed feeding guilds separately in showing that members of two phyllostomid bat communities in eastern Brazil were no more evenly placed in ecomorphological space than were random samples of bats drawn from the entire South American fauna. They concluded that the mobile nature of bats, together with the environmental heterogeneity of their study region, reduced the importance of deterministic interactions in structuring the communities they studied. Alternatively the operational scale for biotic interactions in neotropical bat communities could be the entire forested part of the continent. Studies of some insular birds have shown non-random morphological arrangements that have been attributed to competitive interactions (Case and Sidell, 1983; Moulton and Pimm, 1987), possibly because in such situations immigration has not been enough to swamp the effect. Findley (1989) also showed non-random arrangements of some desert rodents in the American Southwest. There, limited mobility of the taxa may also have acted to preserve the non-random structure. In studying size distributions of North American birds and terrestrial mammals, Brown and Maurer (1989) noted that the log-normal pattern became log-uniform as they viewed assemblages on a scale from continent-wide to that of the local community. They invoked competition among similarly sized species at the local level as one explanation. However there is no significant difference in size distributions of bats at the regional as opposed to the single-site scale. The same right-skewed pattern is seen at both levels (Fig. 7.6). For continental bat communities, then, the pervasive packing pattern, seen at both regional and local scales, seems most likely to owe its origin chiefly to random processes.

Are bat communities saturated?

A saturated community is one which contains as many species as the habitat will support. Wiens (1989) cites various lines of evidence that have led many ornithologists to conclude that most bird communities are saturated. He also noted that at least some of these conclusions may be dependent upon the way in which the data are presented.

Bat communities in north temperate forest zones resemble each other in species richness, supporting 8–16 species depending upon the proximity of hibernacula. This observation suggests saturation for these habitats. However tropical forest habitats in different biogeographic regions support very different numbers of species as noted in this chapter. These differences, as we have seen, are strongly correlated with differences in area and history, suggesting that at least the Old World tropical forests could support more species. But that suggestion carries the assumption that Old and New World tropical forests are alike in their ability to support bats. Some differences between African and South American tropical regions are dealt with by Meggers et al. (1973). Rain forest Africa is seen to be depauperate by any biological measure, presumably including measures of food resources available to bats. The

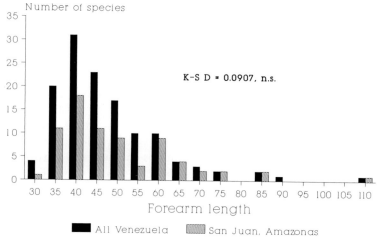

Fig. 7.6. Size distribution of bats of Venezuela compared with that of the bats taken at a single site in Venezuela, San Juan de Manapiare. There is no significant difference between the distributions based on a Kolmogorov–Smirnov test, and no evidence that coexisting species are spaced more evenly than expected by chance.

poverty of African tropical biotas may be attributed to the same factors of area and history that are correlated with species richness of bats. Old World tropical bat communities might be saturated to the level of the capability of their habitats to support them, but there is no compelling evidence one way or the other.

Summary

Studies of biodiversity traditionally cite latitude as an over-riding correlate. However, at least for bats, there is an extremely important longitudinal component as well. Differing ages of the habitats, or differing durations of occupancy by bats do not offer an explanation. Rather, differing areas of tropical forest habitat and differing numbers of putative refuges, either of Pleistocene or of modern age, seem to be the determining forces. Among the tropical rain forest regions, essentially all variation in species richness is accounted for by these two factors. Community, or alpha, diversity is mostly a function of regional, or gamma, diversity, and is thus largely explained by history and geography.

The packing pattern displayed by bat communities and faunas seems most likely to owe its origin to stochastic events.

8 · *Summary and outlook*

Searches for pattern in vertebrate communities encounter the problem that the group selected for study falls short in some way of those characteristics which would be most likely to result in the development of a community structure of the kind expected to occur if conventional models of community evolution are accurate approximations of reality. For example the failure of Wiens and Rotenberry (1980) to demonstrate expected patterns in the shrub–steppe bird communities of temperate North America could have resulted because some of the birds are not permanent residents, and because high environmental variability militates against the development of patterns dependent upon resource limitation and competition. Perhaps the expected patterns could be revealed by studying stable communities of organisms living in stable environments. But what are these organisms? Among small vertebrates, at least, most of the groups that are numerous and readily studied are short-lived species with rapid population turnover and populations that are variable in density. Such r-selected kinds are not ideal subjects for the student of community regularity. Bats are exceptional animals among small vertebrates, in that, although they are small, they are long-lived, slowly reproducing organisms that exist in reasonably stable densities. Furthermore, most bats are tropical, and the greatest diversities occur in tropical forest regions, places which, among terrestrial habitats, seem to approach most closely the ideal of long-term environmental stability. Thus bats would seem to be ideal subjects for the attention of community ecologists.

A few studies do suggest the existence of competition in bat communities. Many reveal differentiation among related complexes of species, with feeding, roosting or other vital activities taking place at different times or in different ways. Bats are often limited by the availability of crucial resources. But it is in their relationships with these resources, rather than with each other, that bats seem most constrained in their community organization.

Despite their suitable biological characteristics, for a variety of reasons bats are among the most difficult small vertebrates to study. Almost every aspect of their behavior, physiology, and ecology, requires rather heroic efforts to document. Moreover, identification of bats is always difficult, even for those with long experience, and frequently requires the preservation of specimens. Thus, although bats seem to be evolved for existence in stable communities, and although they fairly swarm in the world's most stable terrestrial ecosystems, they have proven difficult subjects from which to extract information through the usual methods of ecological study. Partly because of these problems some bat biologists have turned to systematic, morphological, and biogeographic approaches to bat ecology.

Such studies reveal that bat morphology provides a rather accurate key to the kinds of lives the bats lead. Coupled with this insight, world-wide studies of taxonomic and trophic diversity among bats suggest that the greatest ecological diversification is to be found in the Neotropical region and that, perhaps not surprisingly, species richness is an acceptable index not only to ecological diversity but probably to density and biomass as well.

In tropical forest regions, species richness, together with its correlates, at the level of the local community is a function of species richness in the biogeographic region, and that, in turn, is related to area of habitat in the region and to its probable Pleistocene history. Thus, in the huge neotropical rain forest region, species richness is the highest in the world, and local communities, in turn, display the highest known alpha diversities. Little community richness seems to be determined at the local level by local community processes operating on an ecological time scale, but rather it seems to depend mostly upon factors operating at the regional scale over evolutionary time.

Competitive interactions between close relatives may play a part in the microstructure of some bat communities. However, ecomorphological community structure, as depicted by morphograms and distributions of trophic or morphological variables, seems dominated by random processes. A bat community in a large diverse region such as the neotropics is largely a product of processes that have played out over great expanses of space and time, producing a diversity of species whose high motility has brought them together at a single locus within the purview of the contemporary ecologist. Despite our very reasonable expectations, bats seem to be organized into communities structured largely by responses to resources, and to historic and geographic factors, rather than by interactions with each other.

Patterns and their correlations in bats, clear as some of them may be, are of limited interest if they are taxon specific. If some of these patterns are to be found in other tropical organisms, this generality makes them of much greater importance for the tropical ecologist. To what extent are chiropteran patterns in species richness, taxonomic, trophic, and morphological diversity seen in other groups of animals and plants?

Plants

Africa has long been recognized as being depauperate floristically compared with Neotropica and Indo-Malesia. Not only do plants in general, but also angiosperms specifically show this pattern (Richards, 1973). Moreover Richards comments upon the reduced variety of life forms in African rain forest, especially of lianes and epiphytes. Various families of typically tropical plants show the pattern in species richness as well. Palms are abundant in Neotropica (1140 species), scarce in Africa (50 species), and intermediate in the Austro-Asian region (1150 species for the combined Oriental and Australian regions). Orchids show a similar pattern.

Birds

Forest bird species are more diverse in Neotropica than in rain forest Africa. Amadon (1973) compares the avifaunas of the Amazon and Congo drainages, listing 592 species in the former and 212 in the latter region. Borneo alone has over 440 species, a large fraction of them forest-dwelling, so that forest birds of the Oriental region undoubtedly approximate in numbers those for Neotropica as a whole. Karr (1976) makes more detailed comparisons of forest birds in Panama and Liberia, documenting the greater richness of the Panamanian avifauna (51 resident forest species in his Panamanian site versus 41 for Liberia). For study sites with comparable degrees of habitat complexity, Karr showed greater alpha as well as beta diversity in Panama. Both Liberian and Panamanian birds showed the familiar right-skewed size distribution. Karr and James (1975) compared these avifaunas using an ecomorphological approach. Both tropical avifaunas showed increased ecomorphological space compared with forest birds of Illinois, but evidence on packing was equivocal: depending upon the suite of traits used, packing may or may not increase with species-richness. Morphological space was shown to be larger in Panama, a result of greater morphological diversity of Panamanian birds. Tropical forest birds generally seem to resemble bats in the patterns of interest.

Rodents

Emmons (1990) provides a guide to neotropical rain forest mammals, those occurring below 1000 m. One hundred and sixty rodents fall into this category. Using Walker (1975) and Honacki, Kinman and Koeppl, (1982) I attempted to identify tropical rain forest rodents in the Ethiopian, Oriental, and Australian regions. Africa supports 44 species, the Oriental region about 190, and the Australian region 80. Forest rodent taxonomic, morphological, and trophic diversity is by far the greatest in the Neotropics, and least in Africa. Much of the richness of the Oriental and Australian regions results from a profusion of rat-like murids, many closely related to *Rattus*. In a similar way the richness of the bat faunas in these regions is attributable partly to the speciose genera *Pteropus*, *Rhinolophus*, *Hipposideros*, *Myotis*, and *Pipistrellus*. These genera increase species-richness, but do not add significantly to morphological and trophic diversity. Thus, although species-rich, the Austro-Asian tropics are not equal to the neotropics in morphological and trophic richness in either rodents or bats.

Freshwater fish

Tropical freshwater fishes are not so directly dependent upon forest as forest bats, birds, and rodents, but still, a comparison of fish faunas in the major tropical regions should reveal the effects of major processes if such have indeed shaped the tropical biotas. Comparisons here omit the Australian region, as freshwater fish there seem derived independently from marine forms. An old, but still useful source of data on fish faunas is Darlington (1965), while a modern summary of taxonomy and distribution of fishes of the world is provided by Nelson (1984). Even allowing for a certain unavoidable vagueness in estimates from various parts of the world, it is clear that, once more, Neotropica is safely and substantially in the lead with approximately 2500 species. Tropical Africa supports about 1230, but this includes many from beyond the drainages within the rain-forested regions, for example the species swarms from the African Great Lakes, (over 500 cichlids from Lake Malawi, Victoria, and Tanganyika alone). The count from the habitat of interest is probably well below 1000. The Oriental region, as with other groups, is intermediate, with about 1200 species. Thus the general species-richness pattern is again approximately as we might expect if we take our bearings from the other groups considered.

It seems likely that most major groups of forest organisms show the same species-richness pattern described for bats. It is a general, well-supported pattern suggesting that a few major processes have produced it. Just as area and refuge history seem adequate to account for the pattern in bats, they probably account for most of the more general pattern.

It seems likely, as well, that taxonomic, morphological, and trophic diversity follow species richness, and much of the available evidence seems to support that relationship. But that is a matter that can easily be tested. Most intriguing is the possibility that alpha diversity, as well as numerical abundance of individuals and biomass are also correlates of species richness. Here again, a rather straightforward program of study can test this hypothesis. Density estimates and estimates of alpha diversity, taken in the context of a spectrum of gamma diversities, will provide the needed insight.

So bats, after all, do not provide the hoped-for support for competition and resource-limitation as dominant factors providing structure to local communities. Rather they suggest to us that species arise, get together, and manage to coexist. The larger the area, the more species coexist in local communities. The intriguing possibility also exists that, with bats, it may be possible to demonstrate that within each latitudinal zone increased species richness itself may be instrumental in enhancing biological productivity. The study of biodiversity merits much more substantive attention than it is currently receiving.

References

Aldridge, H. D. J. N. and Rautenbach, I. L. (1987). Morphology, echolocation and resource partitioning in insectivorous bats. *Journal of Animal Ecology*, **56**, 763–778.

Amadon, D. (1973). Birds of the Congo and Amazon forests: a comparison. In *Tropical forest ecosystems in Africa and South America: a comparative review.*, eds. B. J. Meggers, E. S. Ayensu, and W. D. Duckworth, pp. 267–277. Washington, DC: Smithsonian Institution Press.

Audley-Charles, M. G., Hurley, A. M. and Smith, A. G. (1981). Continental movements in the Mesozoic and Cenozoic. In *Wallace's Line and plate tectonics*, ed. T. C. Whitmore, pp. 9–23. Oxford: Clarendon Press.

Baagoe, H. J. (1987). The Scandinavian bat fauna: adaptive wing morphology and free flight in the field. In *Recent advances in the study of bats*, eds. M. B. Fenton, P. A. Racey, and J. M. V. Rayner, pp. 57–74. Cambridge: Cambridge University Press.

Barbour, R. W. and Davis, W. H. (1969). *Bats of America*. Lexington, KY: University of Kentucky Press.

Barclay, R. M. and Bell, G. P. (1988). Marking and observational techniques. In *Ecological and behavioral methods for the study of bats*, ed. T. H. Kunz, pp. 59–76. Washington, DC: Smithsonian Institution Press.

Ballinger, R. E. (1983). Life history variations. In *Lizard ecology*, eds. R. B. Huey, E. R. Pianka, and T. W. Schoener, pp. 241–260. Cambridge, MA: Harvard University Press.

Beven, S., Connor, E. F. and Beven, K. (1984), Avian biogeography in the Amazon Basin and the biological model of diversification. *Journal of Biogeography*, **11**, 383–399.

Bezem, J. J., Sluiter, J. W. and van Heerdt, P. F. (1960). Population statistics of five species of the bat genus *Myotis* and one of the genus *Rhinolophus* hibernating in the caves of S. Limburg. *Archives Néerlandaises de Zoologie*, **13**, 511–539.

Bezem, J. J., Sluiter, J. W. and van Heerdt, P. F. (1964). Some characteristics of the hibernating locations of various species of bats in South Limburg. I. *Koninkl. Nederl. Akademie van Wetenschappen. Proc. Series C*, **67**, 325–336.

Black, H. L. (1974). A north temperate bat community: structure and prey populations. *Journal of Mammalogy*, **55**, 138–157.

Bogdanowicz, W. (1983). Community structure and interspecific interactions in bats hibernating in Poznan. *Acta Theriologica*, **28**, 357–370.

Bonaccorso, F. J. (1979). Foraging and reproductive ecology in a Panamanian bat community. *Bulletin of the Florida State Museum, Biological Sciences*, **24**, 359–408.

Bradbury, J. W. and Vehrencamp, S. L. (1977). Social organization and foraging in emballonurid bats. IV. Parental investment patterns. *Behavioral Biology and Sociobiology*, **2**, 1–7.

Brosset, A. (1961). L'hibernation chez les chiroptères tropicaux. *Mammalia*, **25**, 413–452.

Brosset, A. (1966). Les chiroptères du Haut-Ivindo (Gabon). *Biologica Gabonica* II(1), 47–86.

Brown, J. H. and Maurer, B. A. (1989). Macroecology: the division of food and space among species on continents. *Science*, **243**, 1145–1150.

Brown, W. L., Jr., and Wilson, E. O. (1956). Character displacement. *Systematic Zoology*, **7**, 49–64.

Calder, W. A. III (1984). *Size, function, and life history*. Cambridge, MA: Harvard University Press.

Carpenter, R. E. (1969). Structure and function of the kidney and water balance of desert bats. *Physiological Zoology*, **42**, 288–302.

Case, T. J. and Sidell, R. (1983). Pattern and chance in the structure of model and natural communities. *Evolution*, **37**, 832–849.

Clements, F. E. (1916). Plant succession: analysis of the development of vegetation. *Carnegie Institute of Washington Publication* No. 242.

Cockrum, E. L. (1969). Migration of the guano bat *Tadarida brasiliensis*. In *Contributions in Mammalogy*, ed. J. K. Jones, Jr *Miscellaneous Publication University of Kansas Museum of Natural History*, No. 51, pp. 303–336. Lawrence, KS: University of Kansas.

Cody, M. L. (1971). Biological aspects of reproduction. In *Avian biology I*, eds. D. S. Farner and J. R. King, pp. 462–512. New York: Academic Press.

Connell, J. H. (1983). On the prevalence and relative importance of interspecific competition: evidence from field experiments. *American Naturalist*, **122**, 661–696.

Constantine, D. (1966). Ecological observations on lasiurine bats in Iowa. *Journal of Mammalogy*, **47**, 34–41.

Crome, F. H. J. and G. C. Richards. (1988). Bats and gaps: microchiropteran community structure in a Queensland rain forest. *Ecology*, **69**, 1960–1969.

Daan, S. and Wichers, H. J. (1968). Habitat selection of bats hibernating in a limestone cave. *Zeitschrift für Saugetierkunde*, **33**, 262–287.

Darlington, P. J. Jr. (1965). Zoogeography: the geographical distribution of animals. New York: John Wiley & Sons.

Davis, W. H. (1966). Population dynamics of the bat *Pipistrellus subflavus*. *Journal of Mammalogy*, **47**, 383–396.

Diamond, J. (1975). Assembly of species communities. In *Ecology and evolution of communities*, eds. M. L. Cody and J. M. Diamond, pp. 342–444. Cambridge, MA: Belknap Press of Harvard University Press.

Diamond, J. and Case, T. J., eds. (1986). *Community ecology*. New York: Harper and Row, Publishers.

Dobzhansky, T. (1950). Evolution in the tropics. *American Scientist*, **38**, 209–221.

Easterla, D. A. (1973). Ecology of the 18 species of Chiroptera at the Big Bend National Park, Texas. *Northwest Missouri State University Studies*, **34**, 1–165.

Egsbaek, W. and Jensen, B. (1963). Results of bat banding in Denmark. *Vidensk. Medd. fra Dansk naturh. Foren.*, **125**, 269–296.

Eisenberg, J. F. (1981). *The mammalian radiations*. Chicago: University of Chicago Press.

Elton, C. (1927). *Animal ecology*. London: Sidgwick and Jackson.

Emmons, L. H (1990). *Neotropical rainforest mammals*. Chicago: University of Chicago Press.

Fenton, M. B. (1970). Population studies of *Myotis lucifugus* (Chiroptera: Vespertilionidae) in Ontario. *Life Science Contributions, Royal Ontario Museum*, **77**, 1–34.

Fenton, M. B. (1972). The structure of aerial-feeding bat faunas as indicated by ears and wing elements. *Canadian Journal of Zoology*, **50**, 287–296.

Fenton, M. B. (1975). Acuity of echolocation in *Collocalia hirundinacea* (Aves: Apodidae), with comments on the distribution of echolocating swiftlets and molossid bats. *Biotropica*, **7**, 1–7.

Fenton, M. B. (1982a). Echolocation calls and patterns of hunting and habitat use of bats (Microchiroptera) from Chillagoe, North Queensland. *Australian Journal of Zoology*, **30**, 417–425.

Fenton, M. B. (1982b). Echolocation, insect hearing, and the feeding ecology of insectivorous bats. In *Ecology of bats*, ed. T. H. Kunz, pp. 261–285. New York: Plenum Press.

Fenton, M. B. (1985). The feeding behaviour of insectivorous bats: echolocation, foraging strategies, and resource partitioning. *Transvaal Museum Bulletin*, **21**, 5–16.

Fenton, M. B. and Bell, G. P. (1979). Echolocation and feeding behavior in four species of *Myotis* (Chiroptera). *Canadian Journal of Zoology*, **57**, 1271–1277.

Fenton, M. B. and Bell, G. P. (1981). Recognition of species of insectivorous bats by their echolocation calls. *Journal of Mammalogy*, **62**, 233–243.

Fenton, M. B., Boyle, N. G. H., Harrison, T. M. and Oxley, D. J. (1977). Activity patterns, habitat use, and prey selection by some African insectivorous bats. *Biotropica*, **9**, 73–85.

Fenton, M. B. and Fleming, T. H. (1976). Ecological interactions between bats and nocturnal birds. *Biotropica*, **8**, 104–110.

Fenton, M. B. and Morris, D. (1976). Opportunistic feeding by desert bats (*Myotis* spp.). *Canadian Journal of Zoology*, **54**, 526–530.

Findley, J. S. (1972). Phenetic relationships among bats of the genus *Myotis*. *Systematic Zoology*, **21**, 31–52.

Findley, J. S. (1973). Phenetic packing as a measure of faunal diversity. *American Naturalist*, **107**, 580–584.

Findley, J. S. (1976). The structure of bat communities. *American Naturalist*, **110**, 129–139.

Findley, J. S. (1989). Morphological patterns in rodent communities of southwestern North America. In *Patterns in the structure of mammalian communities*, eds. D. W. Morris, B. J. Fox and M. R. Willig, pp. 253–264. Special Publication no. 28, The Museum, Texas Tech University.

Findley, J. S. and Black, H. L. (1983). Morphological and dietary structuring of a Zambian insectivorous bat Community. *Ecology*, **64**, 625–630.

Findley, J. S. and Jones, C. J. (1964). Seasonal distribution of the hoary bat. *Journal of Mammalogy*, **45**, 461–470.

Findley, J. S. and Wilson, D. E. (1974). Observations on the neotropical disk-winged bat, *Thyroptera tricolor* Spix. *Journal of Mammalogy*, **55**, 562–571.

Findley, J. S. and Wilson, D. E. (1982). Ecological significance of chiropteran morphology. In *Ecology of bats*, ed. T. H. Kunz, pp. 243–260. New York: Plenum Publishing Corp.

Findley, J. S. and Wilson, D. E. (1983). Are bats rare in tropical Africa? *Biotropica*, **15**, 299–303.

Fischer, E. A. (1973). The uses and misuses of removal trapping as applied to the investigations of bat communities. *Organization for Tropical Studies course book. Tropical ecology*, **73–2**, 211–221.

Fleming, T. H. (1975). The role of small mammals in tropical ecosystems. In *Small mammals: their productivity and population dynamics*, eds. F. B. Golley, K. Petrusewicz, and L. Ryskowski, pp. 269–298. Cambridge: Cambridge University Press.

Fleming, T. H. (1979). Do tropical frugivores compete for food? *American Zoologist*, **19**, 1157–1172.

Fleming, T. H. (1986). The structure of Neotropical bat communities: a preliminary analysis. *Revista Chilena de Historia Natural*, **59**, 135–150.

Fleming, T. H. (1988). *The short-tailed fruit bat: a study in plant-animal interactions.* Chicago: University of Chicago Press.

Fleming, T. H., Hooper, E. T. and Wilson, D. E. (1972). Three Central American bat communities: structure, reproductive cycles, and movement patterns. *Ecology*, **53**, 555–569.

Francis, C. M. (1988). Composition of the Malaysian rain forest bat community. *Bat Research News*, **29**, 46.

Francis, C. M. (1990). Trophic structure of bat communities in the understorey of lowland dipterocarp rainforest in Malaysia. *Journal of Tropical Biology and Ecology*, in press.

Gaisler, J. (1975). A quantitative study of some populations of bats in Czechoslovakia (Mammalia: Chiroptera). *Acta Scientarum Naturalium Academiae Scientarum Bohemoslovacae, Brno*, **9**, 1–44.

Gaisler, J. (1979). Ecology of bats. In *Ecology of small mammals*, ed. D. M. Stoddart, pp. 281–342. London: Chapman and Hall.

Gaisler, J. and Hanak, V. (1969). Summary of the results of bat banding in Czechoslovakia 1948–1967. *Lynx*, n.s. **10**, 25–34.

Gaisler, J., Hanak, V. and Klima, M. (1956). *Netopyri Ceskoslovenska (Bats of Czechoslovakia).* Prague: Caroline University. (In Czechoslovakian with a German summary)

Gatz, A. J., Jr. (1979). Community organization in fishes as indicated by morphological features. *Ecology*, **60**, 711–718.

Geluso, K. N. (1978). Urine concentrating ability and renal structure of insectivorous bats. *Journal of Mammalogy*, **59**, 312–323.

Gillette, D. D. and Kimbrough, J. D. (1970). Chiropteran mortality. In *About bats*, eds. B. H. Slaughter and D. W. Walton, pp. 262–283. Dallas: Southern Methodist University.

Gleason, H. A. (1926). The individualistic concept of the plant association. *Torrey Botanical Club Bulletin*, **53**, 7–26.

Gould, E. (1977). Echolocation and communication. In *Biology of bats of the New World family Phyllostomatidae*, Part III, eds. R. J. Baker, J. K. Jones, Jr. and D. C.

Carter, pp. 247–279.

Graham, G. L. (1988). Interspecific associations among Peruvian bats at diurnal roosts and roost sites. *Journal of Mammalogy*, **69**, 711–720.

Grinnell, J. (1922). The role of the accidental. *Auk*, **39**, 373–380.

Grubb, P. (1982). Refuges and dispersal in the speciation of African forest mammals. In *Biological diversification in the tropics*, ed. G. T. Prance, pp. 537–553. New York: Columbia University Press.

Haffer, J. (1982). General aspects of the refuge theory. In *Biological diversification in the tropics*, ed. G. T. Prance, pp. 6–24. New York: Columbia University Press.

Handley, C. O., Jr. (1967). Bats of the canopy of an Amazonian forest. *Atas do Simposio sobre a Biota Amazonica*, **5** (Zoologia), 211–215.

Handley, C. O., Jr. (1976). Mammals of the Smithsonian Venezuelan project. *Brigham Young University Science Bulletin, Biological Series*, **20** (5).

Happold, D. C. M. and Happold, M. (1988). Renal form and function in relation to the ecology of bats (Chiroptera) from Malawi, Central Africa. *Journal of Zoology*, **215**, 629–655.

Heideman, P. D. and Heaney, L. R. (1989). Population biology of fruit bats (Pteropidae) in Philippine submontane rainforest. *Journal of Zoology*, **218**, 565–586.

Heller, K. G. and Helversen, O. V. (1989). Resource partitioning of sonar frequency bands in rhinolophid bats. *Oecologia*, **80**, 178–186.

Herreid, C. F. (1964). Bat longevity and metabolic rate. *Experimental Gerontology*, **1**, 1–9.

Hill, J. E. and Smith, J. D. (1984). *Bats. A natural history*. Austin: University of Texas Press.

Holdridge, L. R. (1967). *Life zone ecology*. San Jose, Costa Rica: Tropical Science Center.

Honacki, J. H., Kinman, K. E. and Koeppl, J. W. (1982). *Mammal species of the world: a taxonomic and geographic reference*. Lawrence, KS: Association of Systematics Collections and Allen Press, Inc.

Humphrey, S. R. (1975). Nursery roosts and community diversity of Nearctic bats. *Journal of Mammalogy*, **56**, 321–346.

Humphrey, S. R. and Bonaccorso, F. J. (1979). Population and community ecology. In *Biology of bats of the New World family Phyllostomatidae*, Part III, eds. R. J. Baker, J. K. Jones, Jr. and D. C. Carter, pp.409–441. Special Publication no. 16, The Museum, Lubbock: Texas Tech University.

Humphrey, S. R., Bonaccorso, F. J. and Zinn, T. L. (1983). Guild structure of surface-gleaning bats in Panama. *Ecology*, **64**, 284–294.

Husar, S. L. (1976). Behavioral character displacement: evidence of food partitioning in insectivorous bats. *Journal of Mammalogy*, **57**, 331–338.

Hutchinson, G. E. (1959). Homage to Santa Rosalia, or why are there so many kinds of animals? *American Naturalist*, **93**, 145–159.

James, F. C. and Boecklen, W. J. (1984). Interspecific morphological relationships and the densities of birds. In *Ecological communities: conceptual issues and the evidence*, eds. D. R. Strong Jr., D. Simberloff, L. G. Abele, and A. B. Thistle, pp. 459–490. Princeton, NJ: Princeton University Press.

Janzen, D. H., and Wilson, D. E. (1983). Mammals. In *Costa Rican natural history*, ed. D. H. Janzen. Chicago: University of Chicago Press.

Jones, C. (1965). Ecological distribution and activity periods of bats of the Mogollon Mountains area of New Mexico and adjacent Arizona. *Tulane Studies in Zoology*, **12**, 93–100.

Jones, C. (1971). The bats of Rio Muni, West Africa. *Journal of Mammalogy*, **52**, 121–140.

Jones, C. (1972). Comparative ecology of three pteropid bats in Rio Muni, West Africa. *Journal of Zoology*, **167**, 353–370.

Karr, J. R. (1976). Within- and between habitat avian diversity in Africa and neotropical lowland habitats. *Ecological Monographs*, **46**, 457–481.

Karr, J. R. and James, F. C. (1975). Ecomorphological configurations and convergent evolution. In *Ecology and evolution of communities*, eds. M. L. Cody and J. M. Diamond, pp. 258–291. Cambridge, MA: Belknap Press of Harvard University Press.

Keen, R. and Hitchcock, H. B. (1980). Survival and longevity of the little brown bat (*Myotis lucifugus*) in southeastern Ontario. *Journal of Mammalogy*, **61**, 1–7.

Koopman, K. F. (1970). Zoogeography of bats. In *About bats*, eds. B. H. Slaughter, and D. W. Walton, pp. 29–50. Dallas, TX: Southern Methodist University Press.

Koopman, K. F. (1989). Distributional patterns of Indo–Malayan bats (Mammalia: Chiroptera). *American Museum Novitates*, **No. 2942**.

Kowalski, K. (1953). Material relating to the distribution and ecology of cave bats in Poland. *Fragmenta Faunistica Musei Zoologici Polonici*. **6**, 541–567.

Krzanowski, A. (1956). The bats (Chiroptera) of Pulawy. List of species and biological observations. *Acta Theriologica*, **1**, 87–108.

Krzanowski, A. (1964). Three long flights of bats. *Journal of Mammalogy*, **45**, 152.

Kunz, T. H. (1973). Resource utilization: temporal and spatial components of bat activity in central Iowa. *Journal of Mammalogy*, **54**, 14–32.

Kunz, T. H. (1982). Roosting ecology. In *Ecology of bats*, ed. T. H. Kunz, pp. 1–56. New York: Plenum Press.

Kunz, T. H. (1988). *Ecological and behavioral methods for the study of bats*. Washington, DC: Smithsonian Institution Press.

Lack, D. (1969). Tit niches in two worlds or Homage to Evelyn Hutchinson. *American Naturalist*, **103**, 43–49.

Lagler, K. F., Bardach, J. E. and Miller, R. R. (1962). *Ichthyology*. New York: John Wiley and Sons Inc.

Laurent, R. F. (1973). A parallel survey of equatorial amphibians and reptiles in Africa and South America. In *Tropical forest ecosystems in Africa and South America: a comparative review*, eds. B. J. Meggers, E. S. Ayensu and W. D. Duckworth, pp. 259–266. Washington, DC: Smithsonian Institution Press.

LaVal, R. K., Clawson, R. L., LaVal, M. L. and Caire, W. (1977). Foraging behavior and nocturnal activity patterns of Missouri bats, with emphasis on the endangered species *Myotis grisescens*, and *Myotis sodalis*. *Journal of Mammalogy*, **58**, 592–599.

LaVal, R. K. and Fitch, H. S. (1977). Structure, movements and reproduction in three Costa Rica bat communities. *Occasional Papers, Museum of Natural History, University of Kansas*, **69**, 1–28.

Lehmann, R. (1985). Hunting habits of a northern bat community. *Bat Research News*, **26**, 65.

Leisler, B. and Winkler, H. (1985). Ecomorphology. *Current Ornithology 2*, 155–186.

Lowe-McConnell, R. H. (1987). *Ecological studies in tropical fish communities*. Cambridge: Cambridge University Press.

Ludwig, J. A. and Reynolds, J. F. (1988). *Statistical ecology*. New York: John Wiley & Sons.

Lynch, J. D. (1988). Refugia. In *Analytical biogeography*, eds. A. A. Myers, and P. S. Giller, pp. 311–342. London: Chapman and Hall.

MacArthur, R. H. (1958). Population ecology of some warblers of northeastern coniferous forests. *Ecology*, **39**, 599–619.

MacArthur, R. H. (1962). Some generalized theorems of natural selection. *Proceedings of the National Academy of Sciences, USA*, **231**, 123–138.

MacArthur, R. H. (1972a). *Geographical ecology*. New York: Harper and Row, Publishers.

MacArthur, R. H. and Levins, R. (1967). The limiting similarity, convergence and divergence of coexisting species. *American Naturalist*, **101**, 377–385.

MacArthur, R. H. and MacArthur, A. T. (1974). On the use of mist nets for population studies of birds. *Proceedings of the National Academy of Sciences, USA*, **71**, 3230–3233.

MacArthur, R. H. and Wilson, E. O. (1967). *The theory of island biogeography*. Princeton, NJ: Princeton University Press.

Marquet, P. A. (1990). Competition between distantly related taxa: three reasons why it is not more often reported. *Revista Chilena de Historia Natural*, **63**, 149–156.

May, R. M. (1981). Patterns in multi-species communities. In *Theoretical ecology*, 2nd edn, ed. R. M. May, pp. 197–227. Sunderland, MA: Sinauer Associates, Inc.

McCracken, G. F. (1987). Genetic structure of bat social groups. In *Recent advances in the study of bats*, eds. M. B. Fenton, P. Racey, and J. M. V. Rayner, pp. 281–298. Cambridge: Cambridge University Press.

McKenzie, N. L. and Rolfe, J. K. (1986). Structure of bat guilds in the Kimberley mangroves, Australia. *Journal of Animal Ecology*, **55**, 401–420.

McNab, B. K. (1971). The structure of tropical bat faunas. *Ecology*, **52**, 352–358.

McNab, B. K. (1982). Evolutionary alternatives in the physiological ecology of bats. In *Ecology of bats*, ed. T. H. Kunz, pp. 151–200. New York: Plenum Publishing Co.

Medway, Lord (1969). *The wild mammals of Malaya*. London: Oxford University Press.

Meggers, B. J., Ayensu, E. S. and Duckworth, W. D. eds. (1973). *Tropical forest ecosystems in Africa and South America: a comparative review*. Washington, DC: Smithsonian Institution Press.

Merriman, C. (1990). Relative capture rates of tympanate and atympanate moths by *Lasiurus cinereus* and *Lasiurus borealis*. *20th North American Symposium on Bat Research, Lincoln, Nebraska*.

Morrison, D. W. (1979). Apparent male defense of tree hollows in the fruit bat, *Artibeus jamaicensis*. *Journal of Mammalogy*, **60**, 11–15.

Mortensen, B. K. (1977). *Multivariate analyses of morphology and foraging strategies of phyllostomatine bats*. Unpublished PhD dissertation, Biology Department, University of New Mexico, Albuquerque.

Mosimann, J. E. and James, F. C. (1979). New statistical methods for allometry with application to Florida red-winged blackbirds. *Evolution*, **33**, 444–459.

Moulton, M. P. and Pimm, S. L. (1987). Morphological assortment in introduced

Hawaiian passerines. *Evolutionary Ecology*, **1**, 41–62.

Moyle, P. B. and Cech, J. J., Jr. (1982). *Fishes: an introduction to ichthyology*. Englewood Cliffs, NJ: Prentice-Hall, Inc.

Mumford, R. E. and Cope, J. B. (1964). Distribution and status of the Chiroptera of Indiana. *American Midland Naturalist*, **72**, 473–489.

Myers, P. and Wetzel, R. M. (1983). Systematics and zoogeography of the bats of the Chaco Boreal. *Miscellaneous Publications, Museum of Zoology, University of Michigan*, **165**.

Nelson, J. S. (1984). *Fishes of the world*, 2nd edn. New York: John Wiley & Sons.

Nyholm, E. S. (1965). Zur okologie von *Myotis mystacinus* (Leisl.) und *M. daubentonii* (Leisl.). *Ann. Zool. Fennica* **2**, 77–123.

Odum, E. P. (1967). The strategy of ecosystem development. *Science*, **164**, 262–270.

Oren, D. C. (1982) Testing the refuge model for South America. In *Biological diversification in the tropics*, ed. G. T. Prance, pp. 601–607. New York: Columbia University Press.

Pearson, O. P., Koford, M. R. and Pearson, A. K. (1952). Reproduction of the lump-nosed bat (*Corynorhinus rafinesquei*) in California. *Journal of Mammalogy*, **33**, 273–320.

Pianka, E. R. (1970). On r- and K-selection. *American Naturalist*, **104**, 592–597.

Poole, R. (1989). Ecologists flirt with chaos. *Science*, **243**, 310–313.

Porter, F. L. 1978. Roosting patterns and social behavior in captive *Carollia perspicillata*. *Journal of Mammalogy*, **59**, 627–630.

Prance, G. T., ed. (1982). *Biological diversification in the tropics*. New York: Columbia University Press.

Price, P. W., Slobodchikoff, C. N. and Gaud, W. S. (1984). *A new ecology: novel approaches to interactive systems*. New York: Wiley.

Rautenbach, I. L., Fenton, M. B. and Braack, L. E. O. (1985). First records of five species of insectivorous bats from the Kruger National Park. *Koedoe*, **28**, 73–80.

Richards, P. (1973). Africa, the "odd man out." in *Tropical forest ecosystems: a comparative review*, eds. B. J. Meggers, E. S. Ayensu, and W. D. Duckworth, pp. 21–26. Washington, DC: Smithsonian Institution Press.

Ricklefs, R. E., Cochran, D. and Pianka, E. R. (1981). A morphological analysis of the structure of communities of lizards in desert habitats. *Ecology*, **62**, 1474–1483.

Ricklefs, R. E. and Cox, G. W. (1977). Morphological similarity and ecological overlap among passerine birds on St. Kitts, British West Indies. *Oikos*, **29**, 60–66.

Ricklefs, R. E. and J. Travis. (1980). A morphological approach to the study of avian community organization. *Auk*, **97**, 321–338.

Robbins, L. W. and Sarich, V. M. (1988). Evolutionary relationships in the family Emballonuridae (Chiroptera). *Journal of Mammalogy*, **69**, 1–13.

Root, R. B. (1967). The niche exploitation pattern of the blue-gray gnatcatcher. *Ecological Monographs*, **37**, 317–350.

Rosevear, D. R. (1965). *The bats of West Africa*. London: The British Museum (Natural History).

Sacher, G. A. (1959). Relation of lifespan to brain weight and body weight in mammals. In *Ciba Foundation colloquium on aging*, vol. 1, ed. G. E. Wolstenholme, pp. 115–141.

Schall, J. R. and Pianka, E. R. (1978). Geographical trends in numbers of species. *Science*, **201**, 679–686.

Schober, W., Haensel, J., Natuschke, G., v. Knorre, D., Stratman, B., Wilhelm, M. and Zimmermann, W. (1971). Zur verbreitung der Fledermause in der D. D. R. (1945–1970). *Nyctalus*, **III**, 50 pp.

Schoener, T. W. (1983). Field experiments on interspecific competition. *American Naturalist*, **122**, 240–285.

Schoener, T. W. (1986). Overview: kinds of ecological communities – ecology becomes pluralistic. In *Community ecology*, eds. J. Diamond and T. J. Case, pp. 467–479. New York: Harper and Row, Publishers.

Schum, M. (1984). Phenetic structure and species richness in North and Central American bat faunas. *Ecology*, **65**, 1315–1324.

Seber, G. A. F. (1982). *The estimation of animal abundance.* New York: Macmillan Publishing Co., Inc.

Simberloff, D. (1983). Competition theory, hypothesis testing, and other community ecological buzzwords. *American Naturalist*, **122**, 626–635.

Smartt, R. A. (1978). A comparison of ecological and morphological overlap in a *Peromyscus* community. *Ecology*, **59**, 216–220.

Sneath, P. H. A. and Sokal, R. R. (1973). Numerical taxonomy. San Francisco: W. H. Freeman & Co.

Sokal, R. R. and Sneath, P. H. A. (1963). *Principals of numerical taxonomy.* San Francisco & London: W. H. Freeman & Co.

Southwood, T. R. F. (1966). *Ecological methods.* London: Methuen and Co.

Stevens, G. C. (1989). The latitudinal gradient in geographical range: how so many species coexist in the tropics. *American Naturalist*, **133**, 240–256.

Stevenson, D. E. and Tuttle, M. D. (1981). Survivorship in the endangered gray bat (*Myotis grisescens*). *Journal of Mammalogy* **62**, 244–257.

Strelkov, P. P. (1969). Migratory and stationary bats (Chiroptera) of the European part of the Soviet Union. *Acta Zoologica Cracoviensis*, **14**, 393–439.

Studier, E. H., Wisniewski, S. J., Feldman, A. T., Dapson, R. W., Boyd, B. C. and Wilson, D. E. (1983). Kidney structure in Neotropical bats. *Journal of Mammalogy*, **64**, 445–452.

Thomas, D. W. (1985). Apparent competition between two species of West African bats. *Bat Research News*, **26**, 73.

Thomas, D. W. and Bell, G. P. (1986). Thermoregulatory strategies and the distributions of bats along climatic gradients. *Bat Research News*, **27**, 39.

Toschi, A. and Lanza, B. (1959). *Mammalia.* Fauna d'Italia, vol. IV. Bologna: Edizioni Calderini.

Tuttle, M. D. and Stevenson, D. (1982). Growth and survival of bats. In *Ecology of bats*, ed. T. H. Kunz, pp. 105–150. New York: Plenum Publishing Corp.

Van Tyne, J. (1933). The trammel net as a means of collecting bats. *Journal of Mammalogy*, **14**, 145–146.

Van Valen, L. (1965). Morphological variation and the width of the ecological niche. *American Naturalist*, **100**, 377–389.

Van Valen, L. (1973). Body size and numbers of plants and animals. *Evolution*, 27, 27–35.

Verschuren, J. (1957). Ecologie, biologie et systematique des chiroptères. *Exploration du Parc National de la Garamba*, **7**, 473 pp. Brussels.

Walker, D. (1982). Speculations on the origin and evolution of the Sunda-Sahul rain forests. In *Biological diversification in the tropics*, ed. G. T. Prance, pp. 554–575. New

York: Columbia University Press.

Walker, E. P. (1975). *Mammals of the world.* 3rd edn. Baltimore, MD: The Johns Hopkins University Press.

Warner, R. M. (1985). Interspecific and temporal dietary variation in an Arizona bat community. *Journal of Mammalogy*, **66**, 45–51.

Welty, J. C. and Baptista, L. (1988). *The life of birds.* 4th edn. New York: Sanders College Publishing.

Whitaker, J. O., Jr. 1988. Food habits analysis of insectivorous bats. In *Ecological and behavioral methods for the study of bats*, ed. T. H. Kunz, pp. 171–190. Washington, DC: Smithsonian Institution Press.

Whitaker, J. O. and Black, H. L. (1976). Food habits of cave bats of Zambia, Africa. *Journal of Mammalogy*, **57**, 198–204.

Whittaker, R. H. (1956). Vegetation of the Great Smoky Mountains. *Ecological Monographs*, **26**, 1–80.

Wiens, J. A. (1977). On competition and variable environments. *American Scientist*, **65**, 590–97.

Wiens, J. A. (1989). *The ecology of bird communities*, vols. 1 and 2. Cambridge: Cambridge University Press.

Wiens, J. A. and Rotenberry, J. T. (1980). Patterns of morphology and ecology in grassland and shrubsteppe bird populations. *Ecological Monographs*, **50**, 287–308.

Wilkinson, G. S. 1987. Altruism and cooperation in bats. In *Recent advances in the study of bats*, eds M. B. Fenton, P. Racey and J. M. V. Rayner, pp. 299–323. Cambridge: Cambridge University Press.

Willig, M. R. (1983). Composition, microgeographic variation, and sexual dimorphism in Caatinga and Cerrado bat communities from northeastern Brazil. *Bulletin of the Carnegie Museum of Natural History*, **23**. Pittsburgh.

Willig, M. R. (1986). Bat community structure in South America: a tenacious chimera. *Revista Chilena de Historia Natural*, **59**, 151–168.

Willig, M. R. and Selcer, K. W. (1989). Bat species density gradients in the New World: a statistical assessment. *Journal of Biogeography*, **16**, 189–195.

Willig, M. R. and Moulton, M. P. (1989). The role of stochastic and deterministic processes in structuring neotropical bat communities. *Journal of Mammalogy*, **70**, 323–329.

Wilson, D. E. (1973). Bat faunas: a trophic comparison. *Systematic Zoology*, **22**, 14–29.

Wilson, D. E. (1979). Reproductive patterns. In *Biology of bats of the New World family Phyllostomatidae, Part III*, eds. R. J. Baker, J. K. Jones, Jr. and D. C. Carter, pp. 317–378. Special Publication No. 16, The Museum, Lubbock: Texas Tech Press.

Wilson, D. E. (1983). Checklist of mammals. In *Costa Rican natural history*, ed. D. H. Janzen, pp. 443–447. Chicago: University of Chicago Press.

Wilson, D. E. and Tyson, E. L. (1970). Longevity records for *Artibeus jamaicensis* and *Myotis nigricans*. *Journal of Mammalogy*, **51**, 203.

Wilson, J. W., III. (1974). Analytical zoogeography of North American mammals. *Evolution*, **28**, 124–140.

Zippin, C. (1956). An evaluation of the removal method of estimating animal populations. *Biometrics*, **12**, 163–189.

Author index

Subject index